Black Moon Lilith

M. Kelley Hunter

ISBN-10: 0-86690-611-8
ISBN-13: 978-0-86690-611-1

Cover Design: Jack Cipolla
Cover Photo: Jim Eaton

Published by:
American Federation of Astrologers, Inc.
6535 S. Rural Road
Tempe, AZ 85283

www.astrologers.com

Printed in the United States of America

Contents

Acknowledgements

Heart-felt gratitude to all friends, colleagues and clients who have shared their Lilith stories and insights. This is where the living research happens, between heart and soul and in the trilling of the owls.

Special thanks to:

Maritha Pottenger for soliciting and editing the first small booklet version of this work for Astro Communications Services, Inc., and to Anne Attal for the graphic of Lilith that teased on that cover.

Jim Eaton for sharing his photography with me over the years. The cover image of this book has accompanied me along many deep journeys.

Kirk Reinert and Lilli Farrell for their beautiful mythic visionary art, and generous permission to show "Celestial Friends" and "Weaver of Dreams."

Jan Collins Selman for her awesome images in "The Garden Series" and especially for permission to show "A New Garden for Lilith."

Dana Hunt, for her long sister-friendship and the fun we have creating ideas and images together.

New FaceBook Lilith friends who gave late-night feedback to this work.

Kris Brandt Riske, of AFA, for taking on Lilith, never a simple project!

Finally, to the One who weaves the intricate web of love and life, protecting, guiding, giving and inspiring all sorts of serendipitous, synchronistic discoveries. Is that you, Lilith, expressing, and the guru of my soul?

Illustrations

Introduction

Mine eye gleams on thee, lit with an alien light.
My lips proclaim mysteries. . . .—George Sterling, *Lilith: A Dramatic Poem*

Why use the Black Moon?

Today, there are so many new options and techniques to incorporate into astrology: thousands of asteroids, remote Trans-Neptunian objects, Centaurs, Uranian planets, galactic phenomena, even new planets! Should we even be using Pluto any more since it has been "demoted"? (Few Western astrologers would deny Pluto its due!)

Not a physical body like a planet, Black Moon is a symbolic astronomical point like the Vertex or the Ascendant. It fills the empty focus of the Moon's elliptical orbit around Earth, a point intimately related to incarnation on Earth. The Black Moon offers access to nuances of experience not easily explained by other factors, suggesting subtle states of consciousness and soul-felt issues that are both deeply personal and healing to our world.

The Black Moon is named Lilith, one of the most ancient female images in myth. Renewed interest in the divine feminine has brought many faces of the goddess into modern awareness. Lilith is one of the most alluring. Yet after five thousand years of smoky rumours, we still don't know who she really is, this archetypal figure of ambiguous reputation. As seductive femme fatale, crib death hag or cosmic goddess, vampire or redeemer, she continues to engage our cultural imagination, giving us plenty of background material to mine for our understanding. Considering her visibility in the arts today, she is more popular than ever. In a culture that has largely repressed the creative essence of the feminine for thousands of years, Lilith shines as a beacon of integrity, autonomy and power.

Today's woman or man with a mythic ear and a feminist eye has to question powerful female figures that are judged demonic. This was the case with the Black Moon, as early interpretations focused on its highly negative manifestations. Without denying the challenges, more researchers are discussing its redemptive qualities.

The twenty-first century dawned with a growing awareness of the divine feminine as a co-creative aspect of life. Lilith adds a profound dimension to this reawakening of female power, engaging women and men in fuller self-awareness and soul expression in life.

Lilith snuck up on me, hinting of her presence through literature and myth. It took me years to recognize Lilith as an astrological component. Indeed, there is little to read in English on the subject. This book, an expanded version of a smaller booklet published in 2000, is offered to help fill that gap.

After setting the Black Moon in its astronomical space, a brief mythic overview gives a good starting point to approach its meaning. Astrology is an essentially mythic language, which maintains relevance by the thoughtful inclusion of meaningful archetypes to enrich our self-understanding as evolving humans. The full power and transformational potential of this intuitive science of the stars comes through working with the imagery and energies of the larger Cosmos, in which we are active participants.

Basic, suggestive interpretations through the signs and houses are offered, with a few hints from real life examples. I know that these sections might be the first ones you want to check out. We always want to know more about ourselves. Please take note that the Black Moon, less than any other astrological factor I know, does not easily fit cookbook recipes. Lilith reveals herself as we develop our skills in self-reflection, helping us to discern subtle nuances of our psychology and behavior and to understand elusive particularities of our experiences.

Examples of well-known persons who have a strong Black Moon signature show different aspects of Lilith's charismatic magic, illustrating how this energy plays out in on the stage of life. A further discussion gives clues about Black Moon interactions with the planets, from the personal to the social and outer, generational levels of the solar system, as well as the Nodes.

This book ends with some creative research which I share to express something about Lilith that is almost impossible to put into words. Black Moon is essentially about embodiment, about the way our soul desires to dance upon Earth.

In this third millennium renaissance, the feminine aspect of the divine is reawakening. Like a

High Priestess or as Anima Mundi, the animating spirit of living Earth, Lilith dowses deep waters of the psyche with her divining rods. It is she who stands at the millennial threshold, as she has ever stood at the gate of the Secret Garden or the Garden of Eden, wielding the flaming sword of truth that cuts through illusions and scares away the weak-hearted. Not everyone is ready to walk into such profoundly unknown dimensions of self and experience. Are you?

Chapter One

What is Black Moon Lilith?

In Astronomy

Astronomically, the Black Moon is the second center of the Moon's elliptical orbit around Earth. Elliptical orbits are due to gravitational effects. An ellipse has two "centers," or focal points, which is different from a perfect circle with just one center. Earth is one focus of the Moon's orbit and Black Moon Lilith is the other. Therefore, it is not a physical body but an abstract, geometrical point, like the Ascendant or the Vertex.

The Black Moon is an integral point in the Earth-Moon relationship, symbolizing a central motivating factor that is subtle yet potent, unseen but felt. Named for Lilith, an ancient dark goddess, the Black Moon taps into an urgent desire of the soul that seeks dramatization in life and will brook no denial. As a second center of reference, she gives a sense of rhythm to the Earth, taking us beyond the subjective personality issues of the Moon into subtle dimensions that are essential to our being on Earth.

The center of gravity between Earth and Moon is inside Earth. As part of the Earth-Moon system, Black Moon is like an energetic vortex intimately bound to the center of Earth. Black Moon and Earth are the two foci or the double center, around which the Moon orbits. Lilith is a twin to the core energy of Earth, its deep heart of crystal fire. This central fire feeds and sustains our Earth and our bodies.

The Black Moon is related to the Moon's Nodes, also non-physical points based on the orbit of the Moon. The Black Moon circles the zodiac in eight years and ten months, a half-cycle of the Moon's Nodes. Like the Moon's Nodes, Black Moon Lilith has both mean and true positions, due to gravitational effects. The difference between them can be significant, even up to thirty degrees! Easier to measure, the mean Black Moon is most often used. Of course Lilith can be mean, but she is

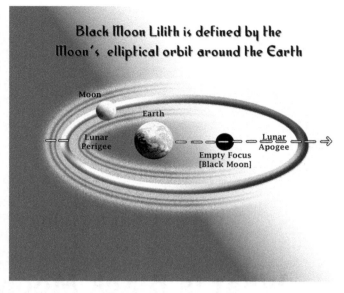

Black Moon Lilith is defined by the Moon's elliptical orbit around the Earth

also true. In her true position, she moves very quickly—up to six degrees per day—and retrogrades often. The true position, sometimes referred to as the oscillating Black Moon, can be ahead of or behind the mean position, sometimes in a neighboring sign. For a fuller discussion of the Black Moon's mathematical complexities, see Appendix.

Closely indicating the Moon's apogee, the farthest the Moon moves from Earth, therefore its closest reach to the Sun, Black Moon is also a reference point in our personal lives that brings us into relationship with the heart of our Sun-fueled experience, an intelligence informed by the wisdom of earthy instinct. Since our culture has lost, even rejected this kind of natural wisdom, as Lilith in mythology was rejected, it is often feared and thus repressed, making it that much more difficult to access and to trust. The Black Moon is where we are pulled into the mystery of life. Lilith insists that we feel through, let go, surrender to something essential and transparent in us that is primal, the passion of the soul. It is a corridor into the heart through which we are lured to discover our deepest truth, the longings and yearnings of our souls.

As we enter a new era in time-space reality, we can use our intuition to guide us, to imagine our way into this new territory, a new mode of being. A wholesome approach involves thinking with the heart and soul, an emerging consciousness directly linked to the current reawakening of the divine feminine. This is the inspiration and aspiration of Black Moon Lilith. In a time such as ours of potent change, Black Moon, with its underlying creative life force and spiritual impulse illumines the inner pathway with the most heart for each seeker of Truth and Love.

The Four Astrological Liliths

Lilith has given her name to several astronomical entities. There are three, even four Liliths. She thus demonstrates the nature of the triple goddess, with a fourth dimension. Before we proceed with Black Moon, here is a brief description of the other Liliths. Confusions abound, typical of Lilith, especially between the Dark and Black Moons. She's not easy to figure out. My book *Living Lilith: Four Dimensions of the Cosmic Feminine* (2009) discusses these components more fully.

Asteroid Lilith

Asteroid 1181, the most solid-bodied Lilith, orbits with most of the other asteroids between Mars and Jupiter. The symbol for asteroid Lilith is a stylized hand, signaling warning, greeting, or blessing. Perhaps the meaning of Lilith's hand signal depends upon how one approaches her. Asteroid Lilith circles the zodiac in four years plus two to four months. In *Asteroid Goddesses*, Demetra George and Douglas Block discuss Lilith in relation to confrontation and issues of equality in relationship. This asteroid specifically questions gender role models.

Dark Moon Lilith

Some say there is a second moon circling Earth—a mysterious dark moon, reported as far back as 1618, reports Delphine Jay in *Interpreting Lilith*, her classic work on Dark Moon Lilith. A second satellite of the Earth, this moon is rarely seen; it reflects no sunlight and can only be sighted on the few rare dates when the sky is dark and it is either in opposition to or crossing the Sun's face in conjunction. It is one-fourth the size of our familiar Moon, say astronomers who have viewed it, and three times as far away. It takes 119 days to orbit Earth, about ten days per sign. Why haven't we heard more about this second Moon? Perhaps it is the nature of Lilith. Its existence has not been completely verified. Perhaps it exists only in an astral plane.

The Lilith Star

The star **Algol** in the constellation Perseus has a reputation as the most evil star in the sky. Most often envisioned as the head or paralyzing eye of Medusa, Algol was also called Lilith by Hebrew star gazers. The star Lilith brings us into a large mythic drama that involves several constellations—Perseus and Andromeda, Cassiopeia and Cepheus, Pegasus and Cetus. The Greek stories upon which our modern Western culture are founded need to be reviewed and revisioned as we enter a new millennium. What is deemed "most evil" may be the key to opening transformational levels of consciousness.

The Dark Goddess Lilith in Myth and Religion

One of the dark goddesses, like Persephone, Hecate or Kali, Lilith expresses the feminine power of the divine, creative life force. Yet these goddess images are most often viewed as fearsome, destructive and devouring. If we follow the mythological trail of these dark goddesses back in time to find the source of their darkness and negativity, we discover not only a major shift in the collective human image of the feminine, but also some deep undercurrent of unease that needs to be acknowledged and healed in our personal lives. Lilith assumes different forms and different names—Lily, Lalita, Layla, Lola, Lolita, Ka-li, Lila—among others. I offer a brief introduction to some of her various appearances in different cultures. Ultimately when we are speaking about the Goddess, all forms are variations of the One. It is part of her play.

In Sumerian Myth

Lilith appears in the oldest known literature, from before 2000 B.C.E. and continues to fascinate us right up to the present day in such venues as the Lilith Fair music fest and characters in popular novels, art and on television. As "handmaid" to the Sumerian goddess Inanna, Lilith brings the men in from the fields for the sacred rites. In another Sumerian story, dramatically retold by cuneiform translator Samuel N. Kramer and storyteller Diane Wolkstein in *Inanna, Queen of Heaven and Earth*, Lilith dwells in the sacred Huluppu tree that Inanna has planted in her garden, the original Garden of Eden.

Queen of Night

A great storm in the deep of the deep had uprooted the tree during the tempestuous mating of two primordial gods. Young Inanna plucks the tree from the water and plants it, intending to use its wood to carve her throne of power and her bed of love. But unwelcome creatures appear in the tree: in the roots, a snake, "who cannot be charmed"; in the branches a fierce Anzu-bird makes a nest for its young; and in the trunk the "dark maid Lilith." The serpent and the bird are regularly associated with Lilith and other goddesses of wisdom and power. The serpent represents Earth-wisdom and oracular powers, and is the symbol for the kundalini energy that awakens and enlightens the chakra centers of consciousness as it courses up the tree of the spinal cord. A winged creature, the bird represents the world of spirit. It opens its wings, enlightened and free to fly up to the Heavens.

Lilith is the conduit from serpent to bird, from Earth to Heaven and Heaven to Earth. These creatures are part of Lilith's untamed nature and have knowledge to give Inanna, who is not yet ready to accept it. She is a young goddess in an increasingly god-oriented pantheon. So Inanna calls in her brother, Gilgamesh, to cut down the tree. The serpent is killed, the bird and its family fly off to the mountains, and Lilith departs for the wilderness. Later, when she is ready for the knowledge of her inner self, Inanna descends to the Underworld and confronts her dark sister, Ereshkigal.

In Hebrew Myth

Lilith is best-known from Hebrew myth, where she was the first wife of Adam. According to one version of the story, Yahweh creates both Adam and Lilith from earth, but with one important difference: he uses impure sediment to make Lilith, whereas Adam is fashioned from pure dust. Therefore Adam expects Lilith to be submissive to him, but claiming equality, she will not be put beneath him. She utters the secret name of God and flies off (on her Sumerian wings perhaps) to lifelong exile near the Red Sea, where she cavorts and consorts with evil spirits. Yahweh sends three angels to bring her back. Though threatened with the death of a hundred of her children per day, she refuses to obey. Meanwhile Yahweh creates another partner for Adam, this time taking one of his ribs while he is asleep and carving it into Eve. She is a part of Adam, created from him and not whole in her own right as was Lilith. This makes it a lot easier for Adam. Myth has it that the jealousy and rage generated from rejection by God and Adam motivate Lilith to come in the night for her revenge, strangling babies and giving men wet dreams to sap their strength. Amulets were worn to ward her off.

In *The Book of Lilith,* Barbara Koltuv shares her research. One story she relates is from the Hebrew mystic tradition: that God made two great lights, the Sun and the Moon, shining with equal brightness. They were not at ease shining together in the same sky. So God said to the Moon, "Go and diminish thyself." The Moon felt humiliated and asked, "Why should I be as one that veileth herself?" Since that time, the Moon has had no light of her own, but reflects the light of the Sun. It is further told that from the demeaned Moon's resentment at the loss of her freedom of choice, Lilith was born—a woman down to the waist and flaming fire or a serpent below. In this form, she is pictured in later Christian iconography offering Eve the apple in the garden of Eden as she winds around the Tree of Knowledge, the Tree of Life.

Another figure in Hebrew mythology is the *Shekhinah*, God's Beloved, known as Sophia in Gnostic Christianity. She is the Wisdom principle, a feminine aspect of divinity, in the Kabalistic Tree of Life. Later Christians came to call her the Holy Ghost. We can imagine Lilith as the "lower" Shekhinah, like the roots of the tree seeking sustenance in the soil, the serpents winding around Inanna's tree, while Sophia is with the bird in the sky-reaching branches and the fruits. As Judeo-Christian religions elevated the masculine aspect of divinity, they de-spiritualized material, sensual reality. The "lower" Shekhinah became unclean, unholy. The Hebrews came into the Sumerian and Babylonian lands, where the Venusian goddesses Inanna, Ishtar and Astarte were celebrated as the embodiment of love in a sacred marriage ritual between priestesses and kings. In these cultures and as well as in the Celtic culture, it was the goddess who gave the king his power through her love and favors. The religious transition to mnotheistic masculine gods made these love rites blasphemous.

As the Great Mother-Lover of the goddess-oriented cultures became split into higher/lower, life-giving/death-bringing aspects, sexuality likewise became split: procreative sex was necessary for life, but sensual sexuality was condemned and demonized. Lilith became an image of denied sexual desire, repressed and projected on to woman, who thereby became the seducer. And yet the sacred love rites did not disappear completely. *The Zohar*, a sacred Hebrew text suggests that the higher Shekhina and the lower Shekhina will be joined in the union of male and female. The sacred marriage rite, the *conjunctio* of heaven and earth as well as of female and male, comes down to us even in the Bible, as the beautifully sensual and mystical *Song of Songs, which is Solomon's*: "I am black, but I am comely."

In *Facing the World with Soul*, depth psychologist Robert Sardello relates a story of Sophia, who is always seeking union with the Spirit of God. The light of God is so brilliant, she moves toward that, attracted to it. Then she is distracted by an equally bright light and finds herself trapped inside the very center of the Earth. Still reaching out toward the divine light, she reaches through every form of life. Sophia incarnates through physical matter, spirit in form, yet at a denser vibration. She is the central flame that is the heart of Earth and sustains the spiritual life of Earth, as the materiality of earth nurtures and sustains our bodies. This flame reaches out from the center of Earth through all living beings, ever longing to merge with the Spirit of God. Found in the darkness, we seek her in the deepest recesses of our Selves.

In Celtic Myth

An intriguing Celtic myth from the Welsh Mabinogion tells a story similar to that of Lilith and Adam. In this story, mother-goddess Arianrhod, for some reason, disowns her sons and places a curse on one of them, Lleu, that he will marry no woman born of the race of men. Lleu is raised by his uncle Gwyddyon who fashions a lovely maiden from flowers and breathes life into her with magic spells. Thus they create Blodeuwedd, a woman not *of* the race of men but rather made *by* them, and marry her to Lleu.

Marital harmony reigns until a hunting party led by Gronw Pebyr, lord of Penlynn, stops by when Lleu is away. Blodeuwedd and Gronw fall madly in love. They cannot part and determine to kill Lleu. This is a tricky business, as he can only be killed under certain unusual conditions that Blodeuwedd persuades him to tell her. The lovers at last succeed in killing Lleu, who turns into an eagle and flies away. However, Gwyddyon finds Lleu and shape-shifts him back to human form. In vengeance, Gwyddyon turns the curse on Blodeuwedd and changes her into an owl, a night bird despised by other birds.

Reflecting the rebellious Eve who ate the apple, Blodeuwedd, awakened to her own desires and passion, rebels against her creators. Though created by males in their own image, she has become self-aware. Facing the choice between continued acquiescence to her prearranged life or following the stirrings of her heart, Blodeuwedd takes her life into her own hands and asserts her will.

This is not an uncommon story even for women today, who follow societal expectations and enter a marriage without passion. In this story, it is as if Eve turns into Lilith. The woman, created by and for man, claims her own power—the power of nature and the nature of womanhood. Blodeuwedd is made of flowers, Natures's beauty at its height. Flowers openly express their sexuality, the epitome of femininity in the natural world. Though not born from a mother's womb, her very flesh is substance of Mother Earth at her softest, most delicate loveliness.

Yet Barbara G. Walker, in *The Woman's Encyclopedia of Myths and Secrets*, reminds us that the Earth goddess has two sides: she is "Virgin Goddess of spring, all made of flower-buds, her beauty disguising a personification of the blood-hungry soil waiting to be fructified with the lifeblood of the sacred king."

The virgin, mother and crone aspects are associated with creation, sex and death. The increasingly autonomous masculine gods, who usurped the power of creation, feared retaliation by the destructive aspect of the Great Goddess, which was connected with the power of female sexuality. In relegating female sexual freedom to the banished realm of devils and despicable creatures such as serpents and owls, the god-dominant culture cursed even the blessings of the awesome all-giver goddess of life and death.

Doesn't this story of Blodeuwedd remind you of Pandora? Also created by and for man, Pan-Dora means "Giver of All." Not only of the ills that Pandora's "box" was said to hold, but more truly a cornucopia of all life's abundance, pleasures and sorrows that we experience in the Great Round of life, death and rebirth. Flower, box, cornucopia, holy grail = the womb. When the birth-giving womb of woman is claimed by male dominion, the free powerful woman who expresses her sexuality is demonized, turned into an owl, exiled, like Lilith. Lilith takes her revenge as child killer, crib death hag and abortion goddess. It has been said that Lilith takes children out of this dreadful world and back to God.

In Hindu Myth

In Hindu mythology, Kali-Lalita is one expression of the dual nature of the Great Goddess that resonates with Lilith. Both a gruesome, ugly death hag and a lovely, flower-bedecked mother/lover goddess, blue or black-faced, she is one of the most beloved and feared goddesses in Hindu culture.

Kali is wild-haired, bloody-mouthed and near-naked, wearing a necklace of skulls and a skirt of severed arms. She may be dancing on or copulating with her consort Shiva's prone body. She is the force that gives life to his inert impulse. She often holds a severed head, sometimes her own, the sacrifice required. What she offers is beyond the "head," the intellect, the mind, the ego. She is the original *shakti*, the divine energy which is the foundation of everything projected from ultimate Reality. She is Nature itself, creator, preserver, and destroyer of universes, who gives form and limits to consciousness, but is herself pure ecstasy, beyond all judgment of positive or negative.

She ceaselessly seeks to reform reality so that higher levels of consciousness are projected into matter. Her creation is ever-changing; she rules over time. The closer we come to Kali, the deeper we feel an intense longing to unite with the absolute. She is the source of all desires and will grant what we ask for. She wrote the Law of Attraction. We don't always realize consciously what we really desire; our actions are often driven by unconscious motivations. Do we seek material satisfaction and pleasure, awareness and power, or do we seek God, the source of all, the Great Mother herself? She will show us what we really want. She is the epitome our our soul desires.

Lalita is a bright side of Kali. Beautiful, sensuous and Self-willed, she, too, is called Mother of the Universe, the life force of the cosmos. She manifests herself in all life, but her true form is seldom seen. It is far too bright for mortal eyes, like the radiance of one thousand rising suns. Robed in blood-red, the color of ripe fruit, Lalita has four arms, which hold her spiritual attributes: a bow made of sugarcane, five arrows made of long-stemmed flowers, a goad and a noose. The bow is the mind which directs the flight of the five arrows of our senses. When our thoughts are directed to God, our senses serve spirit and life is sweet and fragrant. But when our minds become undisciplined, we may lose our aim and stray from spirit. Then Lalita nudges us with the goad. If we still resist her, then she throws the noose around our necks and pulls us back. Even Shiva, Lord of the Dance, surrounded by flames of spiritual power, could not resist her. Her true consort, Shiva recognizes and honors this Great Goddess who stands as the ultimate embodiment of Love, attraction, sexuality—the Life Force.

Derived from Hindu Tantric tradition, a revival of goddess appreciation in the middle of the first millennium, *dakini* came to symbolize and embody the cosmic feminine in Tibetan Buddhism. The transformational power of dakini can come through personal encounters, pathways in the subtle body or spiritually inspiring dreams and imagery. An ego-buster, this dakini shakti sky dancer does whatever it takes to blow your mind open to the spaciousness of pure spirit, the primordial source—like the Black Moon.

The Black Madonna

In Christianity, the healing, redemptive heart of the Black Madonna shows another face of the Black Moon, as Dark Mother in Catholic cults devoted to her around the world. There are over two hundred shrines in Western Europe alone. That may be why European astrologers have been more commonly using the Black Moon. Polish Pope John-Paul II had a deep reverence for his nation's famous symbol, the Black Madonna of Czestochowa, as did Lech Walesa, the once-jailed union worker who became president of Poland. In Costa Rica a basilica was built over the site where a miraculous black rock in the shape of the Virgin Mary was found. She is now the patron saint of that country that has no standing army.

Each Black Madonna, wherever she may be, has a specific history that binds her to the people of the area in a unique way. The traditions that surrounds her appearance or building of her sanctuary, the miracles attributed to her grace and compassion, demonstrate the imminent presence of the Divine Mother who is intimately involved with those that love her.

Lilith in the Arts

Lilith has intrigued many artists, writers and dramatists, including Michelangelo, Victor Hugo, Dante Gabriel Rossetti, John Keats, Thornton Wilder, George Bernard Shaw and Anais Nin. She is a character on the television show *Cheers* and gives her name to the women's music-fest Lilith Fair. More and more references to Lilith come up on internet searches and in new book titles, mostly fiction. There are several web sites dedicated to Lilith, both academic and mythic studies as well as lines of sexy products and services. She has been an inspiration to many in their work with the feminine spirit. We see Lilith associated again and again with secret mysteries, magic, sexuality, power issues in relationship, and with trees. A few examples:

Lilith appears as a main character in *The Shape-Changer's Wife* by Suzanne Schinn, set in the days of Druid magic. Lilith is a solemn woman with an unique, impersonal, still presence. As the story unfolds, the fascinated sorcerer's apprentice guesses the secret, that her magician husband has shape-shifted her into a woman in order to possess her. Her true form is a willow tree in a sacred grove. The apprentice, too, falls in love with her. His love, impelled by a desire to serve her happiness, accelerates his learning to become a master shape-shifter in order to overpower her husband and allow her to return to her true form—a Tree of Life.

A 1960s movie based on J.R. Salamanca's book, *Lilith*, is a fascinating study. Lilith is portrayed by Jean Seberg as a mental patient at Poplar Lodge Sanitarium (there's the tree again). A male patient (Peter Fonda), who loves her, challenges the intern (Warren Beatty), also in love with her: "Do you think they can cure Lilith? You know what she wants. Think they can cure this fire? You know what they have to cure: she wants to leave the mark of her desire on every living creature in the world. If she were Caesar, she'd do it with her sword. If she were a poet, she'd do it with words. She's Lilith. She has to do it with her body."

In the occult fiction novel, *Lilith*, by George MacDonald, she takes the form of a large cat. Mac Donald follows the most-known myth of Lilith from Hebrew tradition that identifies her as the first wife of Adam. He portrays her as willfully claiming the sole power of creation over the Lord, based on her ability to give birth. She wanted Adam to worship and obey her, and when he would not, she deserted him and took up with the Shadow, who made her Queen of Hell. At night, she turns into a leopardess, whose spots become shadows like black holes that stream out and suck blood from babies.

On the other hand, in a 1907 play, *Die Kinder der Lilith* (*The Children of Lilith*) by Isolde Kurz, God blames the woes of mankind on Adam's betrayal of Lilith, the First Woman, embodying the perfected glory of heaven on Earth. Lilith then turns into the fiery angel who bans Adam and Eve from the garden with the flaming sword of truth .

And in George Bernard Shaw's play, *Back to Methuselah*, the serpent in Eden speaks to Eve in a "strange seductively musical whisper" of Lilith, mother of Adam and Eve. It was Lilith, confides the serpent, who first embodied the feminine power of conception and rebirth. She was alone in the Garden when she first saw a creature die and realized that she must find out how to renew herself, to cast the skin like the serpent. The power of conception belongs to the feminine alone, asserts the serpent. Imagining leads to desire, desire to will, and will to creation. Man must give his desire and his will to the woman to share this power. "I dared everything," relates the serpent, who now teaches Eve as it once taught Lilith. "It was by meditating on Life that I gained the power to do miracles."

The Psycho-spiritual Nature of Lilith

Desire—Denial or Delight

Black Moon is energized from the invisible astral or aetheric level. She is the dark sister of Earth, the one buried deep within the core of Earth as the heart beat of our lives, the basic pulse. She is the "black whole" of Earth. As a focal point of the Moon's orbit, Black Moon Lilith works *in relationship* to the Earth-Moon system. Not a physical body, she psychically opens awareness beyond the personal Moon energies into subtle dimensions equally essential to our lives. She is a reference point in our personal lives that brings us into relationship with the heart of our experience on Earth.

Initially our consciousness is conditioned by inborn patterns, family genes and karmic predispositions (Moon material), also bound by the consensus illusions and cultural paradigms of the times. This is a limited and limiting state of consciousness that binds us until we seek beyond it. Black Moon Lilith is our ally in this search, as her nature is revelation, transformation and redemption. She is the all-powerful Mother/Lover who nurtures the soul of her children by cutting the cords that bind us to limited realities and revealing our true nature and essential being. Her sword of truth is like a bolt of lightning that shocks us into awareness. As our consciousness opens up, so do our desires change, continuing to present their limiting dimensions. Our souls never find satisfaction until we let go of all but the One desire for Her Presence and Grace.

Yet on our way to that one-pointedness, our desires demand satisfaction; that is the state of our biological existence. When you cannot do what you want, what your energy naturally wants to do, there arises an area of cloudiness in the mind, a nervous resistance that redirects, sublimates it or represses the natural flow. These "condemned" desires seek satisfaction before they let go of us, so that our souls are desireless, free to merge with pure spirit.

Lilith cannot exist in an atmosphere of betrayal, guilt, shame, criticism, judgment, self-doubt, or any of the negative subjective self definitions or emotional projections that inhibit the pure expression of love. No negatives can exist in a place of pure spirit that is beyond the ego. When we identify with sin and guilt, Lilith must go into exile and banishment. She cannot exist in that limited space that closes off the true essence of our being, the original Garden. She wields her sword of discrimination and cuts to the quick. She sees the cloud of darkness and pops it like a soap bubble, piercing the illusion. She takes you beyond illusion.

As an energy vortex, Lilith Black Moon tunes into a remote band of frequencies. She is essentially Tantric, transformer to higher cosmic octaves. Thus her impersonal energy opens perceptions, transparent areas of the mind not clouded by the weight of collective judgment, by the limitations and illusions of the laws of time and space. The transparent soul perceive beyond such laws, as it melts in the light of its original source. The mind becomes exceedingly erotic, with an objective attunement to the creative life force that expresses itself through the seemingly miraculous. The more spacious one's perception, the more deeply one can embody Truth and Love-Wisdom on the Earth plane.

Cosmic Mysteries

Dark Mater-Dark Matter

Since so much in the modern world judges and condemns natural desires and instincts, Lilith has a challenge. She laughs, as no one sees what she is really up to. Scientists have not yet discovered her subtle secrets. "I am the substance and the one who has no substance," says Lilith in the words of *The Gnostic Gospels*. She dresses suggestively in veils woven from the infinite immensity of dark matter, the subtle unseen webs of energy that provide a foundational fabric for the mere ten percent of the universe that is visible. Like a black cloth with pictures painted on it, the cloth is the substance, the designs mere decoration.

"It is as though the moon threw down a black body upon earth visible only to the subliminal senses, which thereby became illuminated." In *The Black Goddess and the Unseen Real*, British writer Peter Redgrove uses the word "synaesthesia" for the multi-sensory level of gnosis. In such a "poetic" state of awareness, our every day perceptions are de-rationalized in an enhanced mental-emotional-spiritual openness. Subliminal, "psychic" senses are excited through a sensual attunement to a broader context of data on a more subtle level than our normal daily attention. We "know" things in a different way, through direct objective experience which mirrors our inner being in a symbolic way, and the symbols themselves take on a living significance in our immediate experience. This two-way communication, a kind of fusion between inner and outer or between self and other, becomes holy communion. Thus are symbol and metaphor important in imagining and translating our experiences into consciousness which, in the process, alter it. This is truly "virgin" territory, that can be explored by thinking with the heart and letting the desires of the soul lead the way.

Chapter Two

Interpreting the Black Moon

Looking for Lilith

Lilith is experiential in nature and comes alive through living; words alone are not going to articulate her meaning. Let us approach her with all five senses acute and our psychic antennae turned full on. A proper approach to Lilith also entails shedding our mental preconceptions and psychological overlays in order to enter into her primal experience.

Black Moon touches into super-conscious psycho-spiritual processes. In the chart, it identifies where we experience a subliminal state of continual spiritual "crisis" for having to be confined to an ego identity. It is where personal desires may get in the way of how the universe wants to live out that energy through us; an area of our lives that we cannot make happen, but happens through us. A hidden alchemy is in progress, not easily seen by those close to us or even by ourselves, but sometimes becoming a hallmark of our lives. Simply put, Black Moon Lilith is where we must completely let go. No other option will open the door to her secret garden.

There is no cookbook recipe for the Black Moon (what was it the three witches in Macbeth put in their brew?). Lilith can be interpreted only by asking a lot of personal questions. Intimately involved in the details of your life, Lilith knows just how to push your particular buttons, how to show up in specific ways that only you would recognize, sooner or later. Rather sphinx-like is she. Or like a jaguar that silently stalks you with shamanic intent.

A good way to get to know Lilith is to look back at times in your life when she has visited—at transits of Black Moon to your planets or of planets to your true and mean Black Moons. You can also think back to power-packed times in your life and look up where the Black Moon was then: what sign? what house? any aspects? Maybe it was not a Lilith experience. Lilith can sometimes feel a bit like Pluto or Saturn, Chiron or even Sedna. But if Lilith was implicated, then you'll get some good clues about how she has her way with you.

Because the Black Moon is an aetheric aspect of the Earth-Moon system with its Gaia intelligence, its effects may be experienced in weather systems and planetary energy dynamics. On a personal level this translates to the body's systems and health issues, giving us strong indications of blocked energies and timely reminders to open consciousness to the subtle fields of Lilith influence. Energy medicine, spiritual healing and medical astrology offer promising venues of research in relation to Black Moon experience. The twelve biochemic tissue salts, often included in the homeopathic repertoire, are especially interesting as they are associated with the twelve zodiac signs. The body parts and systems associated with your Black Moon placement may teach you a lot about the energy of Lilith in your life.

Life is inherently sexual. Lilith deeply impacts and expresses through our sexuality, as we dance in the primeval yin-yang rhythm of the creative life force. Indeed, some researchers focus primarily on the sexual dimensions of the Black Moon. Lilith evokes a sexual response from the primordial, cellular, atomic, erotic desire that brings us forth into life. Lilith reveals the tantric fusion of sexuality with spirituality, beyond any moral judgments we try to make. Black Moon engages with the invisible aether, the fifth element said to be the subtle dark matter or dark energy from which the four material elements—fire, earth, air, water—emerge. Astrophysicists are aware of the presence of this subtle field, seeing clues in the movement of galaxies.

Lilith is intimate and imminent, measured in synchrony with the Earth-Moon dance, and deeply embedded in our relationship to *Anima Mundi*, the world soul. We are part of the living energy of Earth as it responds to the all-encompassing field of the Cosmos. As humans we have the capacity to realize this and to respond consciously. Lilith leads the way, opening us beyond personality into our larger cosmic identity. We all have idiosyncrasies and annoying, disturbing or neurotic traits that interfere with our fullest presence in life and relationships. Lilith presses these points, seeking to relieve and release pressure by opening the space for a more fulfilling expression. Lilith doesn't tell us which way to go, but opens the space for the potential to arise.

Mean or True

Most astrologers seem to work with the mean Black Moon, yet many interpreters swear by the true. In my experience, professional and personal, I have found that both work. I'm still sorting out the differences in my own dual sign placements, both located in the ninth house—which may be why I am called to write about the Black Moon.

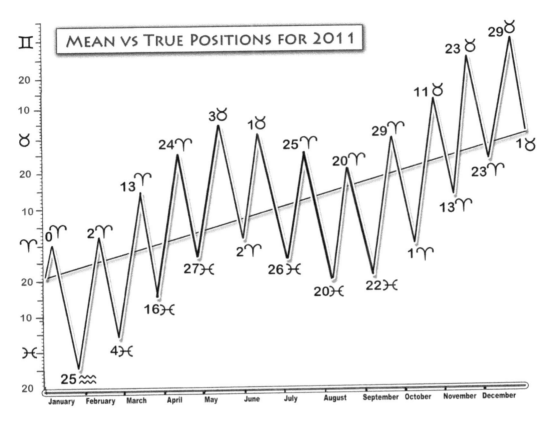

Brad Pitt provides an example to consider. True Black Moon on his Sagittarius Ascendant gives us the Sagittarian cowboy look in *Legends of the Fall* and *The Assassination of Jesse James* by the Coward Robert Ford, but it is possibly his twelfth house mean Black Moon in Scorpio that we see in *Meet Joe Black and Interview with the Vampire*. It seems to be the true Black Moon within a degree of her Leo Ascendant that gives Marilyn Monroe her profoundly affecting aura. Her mean Black Moon is at the end of Leo next to Neptune and on the degree of the star Regulus, the star of royalty. That works too.

Prince William of England gives us a most interesting case of the two Black Moons. This true prince charming and heir to the English throne is a living mythic personage. Born on the Summer Solstice a few hours after a total eclipse of the Sun in Cancer at high noon, he has the mean and true positions very close but hugging the powerful December solstice cusp. His true Black Moon is in the last degree of Sagittarius in the first house two degrees from his Ascendant. Mean Black Moon, just into Capricorn, is opposition his Sun almost to the minute, as well as the strong Moon in Cancer.

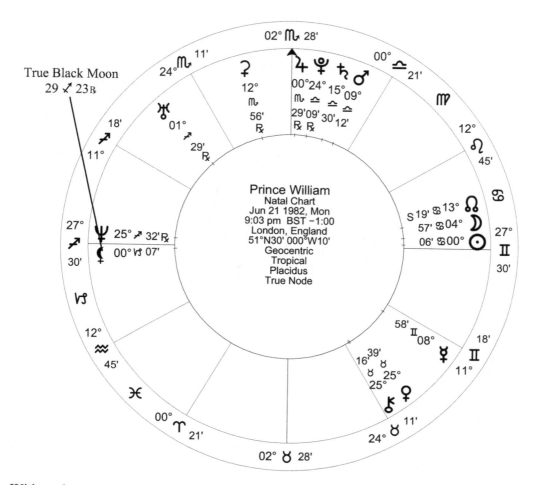

True Black Moon
29 ♐ 23ʀ

Prince William
Natal Chart
Jun 21 1982, Mon
9:03 pm BST –1:00
London, England
51°N30' 000°W10'
Geocentric
Tropical
Placidus
True Node

With such a strong Lilith on the Ascendant, a heightened, sometimes exaggerated sense of self may be hard to live with, being an icon since before he was born and under such constant public scrutiny. He lives a life beyond the "normal" or personal, with a Lilith potential for both an inspired vision channeled through the Sagittarian degree of the galactic center and the true nobleman's sense of responsibility. This young prince was fifteen when his divorced mother, Princess Diana, died in a car crash in a paparazzi frenzy under possibly suspicious circumstances. His father Prince Charles, now remarried to his long time mistress, is first in line to the English throne; William is second. Plenty of Lilith-like scandal among these royals. What will William bring to the royal lineage?

Black Moon Corridor

I propose considering a Black Moon corridor, a kind of sacred garden of the Goddess, using the section of the zodiac encompassed by the true/mean difference in the chart. For some people

Lilith would cover a fairly wide territory, sometimes expressing through two signs, with transiting planets making a longer passage through her realm. Others, with a narrow band of Lilith influence, may have a more focused or intensified experience of her energies, like planets in tight orb.

Black Moon indicates an alchemical process in progress. The mean position describes the overall status of this process; the true position is the ever-moving centerpoint, like the eye of a hurricane. The images of the Sabian Symbols are one way to work with the two positions. You can use both symbols to inspire your Lilith storyline and suggest avenues of expression for the Black Moon.

Black Moon Through the Zodiac

In Fire Signs

Some day after mastering winds, waves, tides and gravity,
we shall harness for God the energies of love and then,
for the second time in the history of the world,
man will have discovered fire.—Teilhard de Chardin, from "On Love"

Galactic Angel

The life force of Black Moon in fire can burn hot, radiating or enlightening in the three fire signs, like different types of stars. In Aries it can be like a newly minted, hot blue-white star; in Leo like the radiant golden Sun; in Sagittarius like a red super-giant star getting ready to explode into a supernova. We see Lilith in nature in wild fires and forest fires, even the underground fire of volcanoes, all the fires that both rage in their destruction yet clear the ground for new growth. The intense heat of fire can lead to burn-out or regeneration. Fire signs are often accused of being selfish. The identity of this small or capital "S" self/Self is the exploratory province of these signs and often their Achilles heel when Black Moon is positioned here.

Aries

She had a mighty will: she strove and strove
and willed and willed for more moons than
there are leaves on all the trees of the
garden.—G.B. Shaw, *Back to Methuselah*

Aries is the electric fire that sparks life out of the black void. Responding to a higher center on an aetheric or causal plane, it conceptualizes impulses from that center, generating energy in a spiral like the Milky Way Galaxy. It creates a new world in a powerful flash of the will-to-be. With Black Moon in Aries these flashes are followed by instant, instinctive action. There is no time for questioning because the action takes places before the conscious thought coalesces. If the lower mind interferes, there can be a kind of paralysis of not being able to take action, a painful, self-conscious block. "It usually works out better if I don't think about it, so my mind doesn't have to try to decide," explains one woman with this placement. This is the path of action, acting on the impulse of cosmic desire To Be, a good strategy when you are following a clear signal.

Aries Black Moon faces the challenge to step out into fresh territory. At Black Moon moments, one is pushed beyond one's comfort level into new territory where there are no markers on the path. The insistence upon maintaining personal integrity is so strong that it may cause rebellion against guidance from others. In relationships, the urge to be one's Self is urgent, often leading to the severance of a partnership in order to pursue the solitudinal quest for Self. Aries Black Moon challenges the ego's assumptions of autonomy and insists that you be who "I AM." Black Moon will not let one compromise very far. Sometimes there is no leeway at all. A Pisces true Black Moon will be more allowing, yet necessitate a stronger flash of fire to clarify the impulse emerging from the diffuse sea of potentiality.

If you have Black Moon in Aries, you may be ready to take a risk at a moment's notice or at any crossroads in life. You rarely choose the safe road. If you get off on the wrong track, you can be accident-prone. Slow down when you start getting such signals. The "mighty will" of Black Moon in Aries needs to be honed like a sharp sword into a tempered tool, a powerful willingness. Otherwise it can run rampant into willfulness, brashness and rashness. Yet when reigned in too tight or denied freedom, raging headaches can be the debilitating effect.

You likely need time alone to do your own thing without any input or imposition from other people. Approach experience with "beginner's mind," as the Buddhists say, every time as if it is the first, with no preconceptions. Black Moon in Aries relishes that freshness. Beginner's mind is not naivetee, although you may exhibit this trait as well. Mindfulness is a state of extraordinary attention and intention, strengthened when accompanied by a Taurus co-placement that desires a sensate manifestation. With a Pisce co-placement, a crystal clarity can arise seemingly instantaneously from a deep emotional undercurrent or psychic tide. Engaging the opposite sign Libra can help balance an over-emphasis on yourself. In the experience of self and other, you can listen, have dialogue, receive feedback. Relax in the face of criticism and remember to listen with a healthy dose of objectivity. The issue may not be about you at all, but the other person. Reactivity will short-circuit any benefit there may be in the interaction. Still, your Black Moon in Aries won't let anyone else tell you what to do!

Beatles front man John Lennon, inventive actor Gary Oldman and Spanish artist Francisco de Goya are good examples of this placement.

Leo

In the bright crystal of your eyes
Show the havoc of fire, show its inspired works
And the paradise of its ashes.—Paul Eluard, *Psychoanalysis of Fire*

Leo is the solar fire that glows in steady, magnetic waves, radiating love-wisdom from the heart of the life source. Black Moon here takes on a dramatic flair, rather like the huge surges of solar nuclear energy that we see from this distance as sun spots. Natives experience the effects of these solar flares and coronal mass ejections (CMEs) in their internal and external weather. The creative life force is embodied in a powerful urge for Self expression. One native says it thus: "I do have 'surges' of powerful energy that tend to blast out, but they are largely behind the scenes and I am not really very liberal about randomly throwing my energy around."

These people are particularly magnetic, with a fascination for the "larger than life." They seek to express and embody archetypal dimensions. Yet the "royal" ego may claim center stage, insisting on casting one in certain roles and usurping the creative power. This constriction may be imposed on oneself, but often is experienced in the form of powerful people or groups in one's life that undermine or inhibit the full expression of one's true essence. The life of Princess Diana demonstrates these themes. She broke the mold of the royal family in a number of ways and revealed herself and her wounds to the world beyond the role she was assigned. She and her lover rebelled and were destroyed. She publicly stated she wanted to be the queen of our hearts.

Black Moon in Leo may give a flair for expressing spiritual ideas in creative ways, an interest in the esoteric arts or artistic expression of esoteric images, the hidden, "dark" side. A former air force officer made a dramatic change of life when his subconscious urges called for a broader expression of aspects of his inner being. He is now a radical gay fairy involved with costuming, ritual and street theater in the context of a group "exiled" from mainstream culture.

The inner nuclear power of the Self shines strongly on others with a challenge to the core self of others to express their fullness. One who has lived through the breaking of one's own heart in seeking the heart of true Love, can hold the mirror steady for others to find the courage to follow the path with most heart for them.

Leo is an emotive sign. Emotions are in motion, as differentiated from feelings, which are in the moment and governed largely by water signs. Emotions are often locked in the body's tissues at highly-charged times of life, creating touchy points where we become emotional and wastefully discharge energy in unconscious actions. The Lion as king or queen of the jungle becomes master of these emotions, taming the heat of anger, frustration and repression into heart-filled generosity of spirit. If the true/mean positions involve Cancer, the distinction between feelings and emotions is the exploration of a lifetime. If one position is in Virgo, the taming of the Lion by the Goddess, as in the tarot card, Strength, places the heart on the alter of service.

If you have Black Moon in Leo it is essential to find some form of creative expression, arena of leadership or other stage on which to shine in order to satisfy Lilith's urge for dramatization. Otherwise you may tend to "act out" in ways that may be less than loveable. Participating in global meditations, communal gatherings or volunteer activity can be another way to offer your heart and pull in the Aquarius polarity, easing any incipient narcissism. The heart center may be your most active chakra. If your Lilith is not allowed full scope, your heart will speak to you through some irregularity. Essential for heart health is self esteem and self love, the deepest lesson of Black Moon in Leo. You realize you can't truly love anyone else until you love yourself, nor can others give you the sense of full self acceptance that can only come from inside. The HeartMath Institute has leading edge teachings that may speak to you.

Christopher Reeve, Marilyn Monroe and Robert DeNiro are all on the list of shining stars with Black Moon in Leo.

Sagittarius

>*With dancing, red-hot sparks about,*
>*All of a sudden her wings spread out,*
>*Dreadfully changed she seems to him,*
>*Kin of the flaming Cherubim.*—Isolde Kurz, *Die Kinder der Lilith*

Sagittarius is fire by friction produced from the interaction of multi-dimensional levels of the self, a revelation of life force on the go. Imagine an atom, with all its particles in motion. This motion ever expands its frame of reference in the space between particles and through photons, motes of light energy that jump in and out of excited states, exchanging energy with other molecules. This is the way an individual consciousness interacts within the universal field.

Sophia, the Holy Ghost of Wisdom, is especially active in this position, giving a kind of prophetic gnosis that recognizes a moment of truth. These people don't beat around the bush. Their urgency for the truth is like a goal they relentlessly pursue, as one level of truth leads to the next. If the quest is reigned in or constrained in limiting constructs, the fire of the life force dies down and forms of depression may ensue. Black Moon in this sign of aspiration reaches ever further into the galactic field of wisdom. The center of our galaxy is located from Earth at the tip of The Archer's arrow, where the "royal road" of the Milky Way is at its star-strewn creamiest.

A challenge from Black Moon in Sagittarius is to "walk one's talk." As Black Moon seeks embodiment, the truth must be lived out and tested. This is not easy for Sagittarian fire, impatient with the time it takes to do the testing and grounding, A true Black Moon in neighboring Capricorn will be more strict about this, but also supportive. A Scorpio-leaning placement can lead to very unusual experiences of an esoteric nature, creating a possible crisis of belief and an urge to speak "the truth" at all cost.

At times certain archetypes or spirits on another level of reality may walk into one's life as guides or teachers. The visionary tendency can be broad and futuristic, like that of Nostradamus with his Black Moon in Sagittarius, or one can feel misunderstood and ignored, like Cassandra and the Boy Who Cried Wolf. The intuition is very strong here, as the *synaesthesia* of multi-sensual perception is enhanced. It is necessary to turn off the mental chatter based on "tapes" from the past in order to give the intuition room to move and open the higher mind to its capacity to know. The Sophianic impulse may lead one into "higher" education, academic and/or esoteric studies.

One woman with Black Moon in Sagittarius has had a life-long interest in spiritualism, which she integrated into college studies and into her work as a psychiatric nurse. Her academic work in psychology has been fueled by erotic dynamics with men in unusual circumstances. She married a man from another country and found herself "exiled" when he had a stroke.

If you have Black Moon in Sagittarius you are a seeker. Inevitably through your life the goal of your search changes as broader contexts of understanding open. Religion may appeal but not satisfy your soul hunger unless you have the freedom to explore the philosophical underpinnings. Spiritual teacher Deepak Chopra, French philosopher Gaston Bachelard and theoretical visionary Noam Chomsky are found among these natives.

The Gemini opposition keeps new pieces of the puzzle coming your way, so you don't settle rigidly into any dogma. It also facilitates the communication of your intuitive awareness. You may have talent for intuitive sciences, "reading" hidden texts and translating esoteric or lost languages. Writing and storytelling are other talents of this placement. You may have a fascination with the "foreign," leading to success abroad or travel toward distant horizons, like the bear who went over the mountain to see what she could see—and kept going and going. There is always another mountain to climb.

In Earth Signs

I am the substance and the one who has no substance.—Gnostic Gospels

As the second focus of the Moon around Earth, Black Moon is a kind of twin to Earth. Perhaps this point should be called Black Earth. Black Moon has a strong affinity with earth signs, in its capacity to *embody*. The telluric qualities of Black Moon in earth signs are powerful, the body's senses alive with psychic currents. Earthquakes that can split the Earth open, the mineral-rich molten lava or toxic suffocating ash that erupts during a volcano, underground caves with stalactites and stalagmites, crystals and geodes—these are Black Moon manifestations that show the power of earth as an element of life.

Taurus

> In the roots of the tree was a serpent,
> in the branches an Anzu bird had made a nest for its young
> and in the trunk, the Dark Maid Lilith.
> —Wolkstein and Kramer, *Inanna, Queen of Heaven and Earth*

A New Garden for Lilith

This is one of the most natural positions for Black Moon, Gaia, Earth goddess, herself: "I, mater, I matter am." This position suggests an archaeological dig of the soul, a bedrock attunement with Earth substance. This is the belly dancer goddess, adorned with flowers, the sexually exuberant beauties of nature. She moves with the core rhythms, feeling the earth vibrations in a slow and deep way. Pachamama, the Amazonian tribes call her in the living richness of the lush, healing rain forest. Taurus is the tree of life, pushing its roots into the mineral-rich soil, down even to Earth's center, so it can reach that high into the heavens and produce fruits of rare quality. The sensual nature is fully explored, so that soul values are embodied over time. "By the fruits of their labors shall you know them," as the Bible says.

Taurus deals with the stuff of life, sometimes burdened by possessions or thick in the thrall of attachment to money, possessions, sex. Over-emphasis on materiality can be a persistent lesson, if not a downright downfall. People with Black Moon in this sign develop great inner strength and are called to harvest inner riches in the fields of the soul, seeking to develop their resources with a resilience that plows through all resistance. They tend the inner as well as the outer garden, honing values that sustain life in a simple, most basic way. Their lifestyles and values may be the hallmark of their lives. Black Moon Taurus contemplates spirit in substance, using spiritual tools to place the pillars of the temple of heaven on earth. Experiential and kinesthetic, these natives can get "down and dirty" and dance their body's wisdom.

A Taurus woman with Black Moon in her sign molds ancient forms of pottery and etches her pieces with petroglyph symbols of soul power. She sculpts human figures in intimate connection with plants and animals. She describes her work as "handmade functional art pottery for the ritual of every day life." Onsite at an archaeological dig in the Caribbean, her fingers pulled out a small detailed stone bead over eight hundred years old. Similarly, the Dark Goddess finds the nugget of the Earth in the vast tapestry of the sky field and declares it to be of value.

If you have Black Moon in Taurus, you are a natural tree-hugger. Getting your hands in and on earth can be most relaxing. Inhaling the loveliness, aroma and blatant sensuality of flowers or grazing on the colors, textures and tastes of vegetables growing in the garden, rooting out the hidden muskiness of wild flowers or mushrooms deep in the woods all bring home the basics of life. The simplicities and ordinariness of life have a sustaining rhythm and grounding. You glory in the textural richness in any endeavor you are engaged in. Buy dishes you love to wash.

For a painter, the smell of the oil colors or the rough ripple of watercolor paper will bring great satisfaction. Only natural bristle brushes will do. For a violinist, it may be the smooth wood and curves of the instrument and the resonance of the bow across the strings next to the ear. A woodworker will lovingly finger the nuances of the grain. Allowing your body to move in the flow of air, to the rhythm of music is most centering. If you are feeling constricted, open your throat and let out the primal sound. Toning, mantra and singing answer the call. Allow Lilith's aetheric Earth pulse to move through your body, your primary instrument.

The Scorpio opposition stimulates the regenerative urges and supports the Black Moon by questioning any basis on which practical Taurus seeks to build security and stability. The soil of your life is periodically turned over, enriched with the compost of wisdom gained from your experiences, even those you consider failures. True Black Moon in either of the neighboring signs will add a restless undertone. A true Black Moon in Aries will ignite fresh ideas, motivating the Taurus to take more risks than it otherwise would. A Gemini influence will spread across the broader field, like a honey bee gathering pollen to bring back to its hive. Listening to different opinions, learning new things, keeping up with the latest information will help Taurus to be more effective and enduring.

We find Russian Empress Catherine the Great, actress Greta Garbo and ground-breaking scientist Kurt Godel among these natives.

Virgo

What is inside of you is what is outside of you
And what you see outside of you, you see inside of you;
It is visible and it is your garment.—Gnostic Gospels

With Black Moon in Virgo, the body responds strongly to telluric energies, the rhythms and pulses of Earth. The connection to nature in general is strong, but there may be a deep resonance with specific realms, be it flora or fauna, weather patterns or the elements. Certain of the body's animal senses can be acute, i.e., smell, hearing, taste, night vision. The skin may be especially sensitive to psycho-sensory currents. Virgo is the most sensually-nuanced sign. The body can be quite particular in its requirements, so that one's physical vehicle, the temple of spirit, remains finely tuned. The veil between the inner and outer worlds is like a filtering membrane that can be quite thin. One can experience acute or chronic environmental sensitivities that necessitate the filtering activity, so one learns to monitor what energies one allows in. Virgo seeks purity, a good concept to contemplate.

This placement contains specialized talents that seek appropriate circumstances in which to manifest. One seeks particular energy fields for the fulfillment of the task at hand. One woman finds that her life circumstances change abruptly in literal ways when she identifies the next lesson that she is to learn. A sign language interpreter with Black Moon conjunction Pluto in Virgo was revived after being pronounced dead in a car crash. The experience left scars in her body that urge her to practice yoga and belly dancing. She has become an advanced BodyTalk practitioner.

The priest or priestess, robed in dark matter, is the Initiatrix in this sign of the Winged Goddess, as one seeks a way to live life as a sacred ritual. Other people may react to Virgo Black Moon in unusual ways because of the interpenetrating nature of energy fields that pass through their bodies. They tend to have an instinctively tantric approach to sexuality. If the mean/true positions lean into Libra, partnerships are key to experiencing this sacredness; if into Leo, the urge for personal expression is put into service of the Divine.

Lilith's bird is most often the screech owl that flies in the dark of the night. Sometimes nicknamed a feathered wild cat, the common screech owl sings songs and duets described as both tremulous and lugubrious. One song is a mellow whistle that ends in a shrill whining or screech, thus its name. The mating duet has a "trilling, cooing sound, with a soft angelic quality," says nature tracker Hunt Harris. With Black Moon right on his Virgo Ascendant, he hears this song often at certain seasons.

If you have Black Moon in Virgo, you need time in Nature as your psychosomatic mind-body vehicle continually tunes into organic rhythms of life and can become stressed by artificial environments and schedules. As the Earth's electromagnetic fields are highly responsive to cosmic input from interstellar space, so is your system. Your body may require organic produce and a refined diet. Virgo rules the intestinal system through which nutrients pass or are eliminated. Virgo uses the sifter function of the membranes. If you want to make an angel cake you need the finest flour.

Consider that your body is an alchemical vessel and pay attention to every ingredient that goes into it. Relax with it, no need to become hypochondriac or hyper-vigilant about it. Your body can take care of itself. We now know that there is intelligence in every cell of the body, not just the brain. Trust that instinctive wisdom and you'll find yourself in the right place at the right time. Engaging in a sport, yoga or qui gong, a dance form or some similar activity can heighten awareness of your physicality. Even while you hone your specialty and the particular form of service your Black Moon requires, take it easy on yourself. The compassion of the Pisces polarity goes a long way toward softening the Virgo tendency toward self-criticism that can undermine confidence. Practice non-attachment to the outcome of your actions. Be like the Persian rug-makers who leave a dangling thread in humility before the perfection of God/dess alone.

The fourteenth Dalai Lama, *Le Petit Prince* author Antoine de Saint-Exupery, and scientist David Bohm all demonstrate sensibilities of this placement.

Capricorn

> *I am Nature, the universal Mother, mistress of all the elements, primordial child*
> *of time, sovereign of all things spiritual, queen of the dead, queen also of the*
> *immortals, the single manifestation of all gods and goddesses that are. . . .*
> —*The Metamorphoses or The Golden Ass* by Apuleis

The Crone slips the Sword of Truth from the stone of old convention. Capricorn Black Moon is a Mistress of Mystery, formulating spiritual principles and forging enduring structures with her innate sense of sacred geometry. She is architect and builder of standing stone circles. She gives the formula to turn lead into gold and draws up blueprints for the atomic structures of crystals. Patiently over time and with natural discipline, she wields the proper pressure to transmute coal into diamond. She is the Sibyl who carved the runes of prophecy. she lives in the neolithic grandmother stones seen the world over.

This placement designates an apprentice to the Dark Goddess, charged by and with her authority to demonstrate her power. When they "graduate" from their apprenticeship through their own particular set of circumstances or formal practice, those with Black Moon in Capricorn are in a position to guide, rule or train others. They often work with the realm of nature spirits to support the capacity of their own bodies to ground energy. They learn that they can't get through life on their own without support from spirit, though their limits of self-sufficiency are tested till there remains not a sentimental bone in their bodies. They have an impact on space-time reality, embodying the eternal spiritual principles and sustaining the life force, often forging new forms in their fields through their own experience. A literary example of this is Anais Nin, who wrote innovative fiction and her famous diaries. An example from science is Marie Curie.

With Black Moon on her Capricorn Ascendant, a practitioner of esoteric and counseling arts carries a strong sense of authority. She develops experiential forums through which she and others can consciously de-structure old patterns and incorporate more effective ones into their lives. When a man experienced too much emotional abuse in his marriage, he left it, thereby becoming a single father. Men as well as women can live out the archetypal role of the Lilith who leaves a controlling marriage. This man continued to integrate the power of his own feminine nature through intensive therapies such as holotropic breathwork and alchemical hypnotherapy which he now practices, often healing people in positions of power.

If you have Black Moon in Capricorn, you are on a path to master your work. Any tradition you are drawn to becomes a container for your abilities to be developed to their full capacity. This is an ambitious sign, not in the worldly sense, but because there is such a sense of responsibility for accomplishing your mission in life. A Sagittarius co-placement will emphasize that sense of mission, sometimes into righteous fervor (not always a bad thing!). Most often the mission involves taking your experience and making a meaningful contribution to the tradition you follow. An Aquarian co-placement further attends to the social relevance of the work. You'll feel it in your bones when you are on the right track. The satisfaction you take in your work is the best reward. You may not gain the worldly recognition you seek. Sometimes the work is deeply internal.

The influence from the Cancer opposition may require you to take care of others and their needs before your own. If such circumstances pull you off your own track, Lilith will let you know in a particularly apt way. Otherwise the care-filled tending of others can forward the mastery of your skills in a significant way. Still, the polarity between dependency and self-sufficiency is an ongoing and complex dynamic on many levels, personal and professional, and deeply embedded in a few key relationships, most likely familial or teacher/student. Emotional maturity is thrust upon you one way or another. Certain situations lead to issues of control and manipulation from either side or both. Capricorn likes to be the one in control. Self control is the ultimate mastery.

In Air Signs

> How do you know but ev'ry Bird
> That cuts the airy way
> Is an immense world of delight,
> Clos'd by your senses five?—William Blake

The Sumerian root of Lilith's name "lil," meaning storm demon or wind spirit, elicits the winged, bird aspect of Lilith flying through rarified atmospheres. Air is the breath of life, the "Word," the power of thought and language. Atmospheric effects in nature range from electrically-charged lightning storms, wild hurricane winds and twisting tornados to the fresh breezes that rustle the leafy trees into whispering oracles. Air circulates around the globe, like migrating birds, spreading dust, ash, pollens, all kinds of information. Black Moon can lift to its highest intelligence in the air signs.

Gemini

Lilith uttered the ineffable name of God and lifted up into the air.—from Jewish legend

This Lilith position can bring out the best of Gemini's native intelligence. In the sign of The Twins, Black Moon plays with duality and paradox within self, in relationship and in the context of the moment. The paradox of dark and light is often exaggerated and can sometimes be experienced in extreme ways. Black Moon in Gemini can feel suffocated or trapped by rationality, breathless for novelty and variety. Restless and continually exploring, one is intrigued by the foreign and "forbidden," in travel, learning and sexuality. Relationships can be more complicated than Gemini likes in the ongoing tennis match between the instinctive and mental aspects of one's nature. Lilith in Gemini can be the *puella* or *puer*, the creative, ever-youthful Peter Pan or Tinkerbell, flying off at the least sign of danger or commitment. A Cancer influence emphasizes the emotional work often needed to overcome the impact of infancy in the desire for bonding.

Celestial Friends

The Gemini tendency toward superficiality results when the soul has no real interest in the activity at hand. Learning "problems" early in life can be the result when the natural intelligence is coerced away from its own way of relating to new information and asking questions. The mind is always making connections even while in the middle of a conversation. Black Moon Lilith smiles like the Cheshire Cat, fascinating and alluring, amused by any hesitancy to surrender to her deeper level of knowing beyond the latest mental framework. She utters magical formulae and prophecies, as she taps directly into an objective inner logic unregulated by the status quo mind-set, which dictates the way our minds are allowed to think. It can be difficult to articulate this inner wisdom in normal language; multi-media may be employed. Gemini Black Moon is often multi-lingual and reads or writes to know how to say what it knows. Nature is the truest "book." There can be an affinity with and skill in working with young people, particularly teenagers.

A woman writer comes from a diverse heritage. A life-long study combining Feminist, Native American and Black history, psychology and spirituality has informed her natural prophetic capacity. "My people tell me what to say," she says, which is whatever a person needs to hear now to go to the next level. Her writings are penned in a unique voice; she has trouble with academic form. Her teachings are most powerful when she speaks them; sometimes she can turn your mind around.

If you have Black Moon in Gemini you can be an excellent and witty communicator. Playful spontaneity can bring the best of you out, so play! Do things for no reason at all. Life is an improvisation. Let the patchwork quilt of your interests and curiosities lead you to rare perceptions and various ways of conveying them. Gemini's flexibility and the influence of the Sagittarian opposition may lead to a gift for teaching because of the unique way you put information together for a particular audience. Trust your high intelligence. You usually don't need to second-guess yourself. Such self-doubting can be hard on your nervous system, that already has a tendency to channel extra kilowatts. A kinesthetic learner, even more so with a Taurus co-placement, which grounds you more solidly into your physicality, you may learn and speak best when moving your body, especially your hands. Find something to do with those restless fingers. Massage, knitting, sign language, or playing an instrument (drums?), etc. can be an interesting exercise on many levels. When in doubt: breathe—deeply. Make sure you are oxygenating your body to calm your excitable nervous system.

Blind educator Helen Keller, tennis pro Rafael Nadal and Wolverine actor Hugh Jackman are some examples with this placement.

Libra

Lilith questioned Adam, "Are we not created equal?"—from Jewish Legend

The Black Moon principle of relationship is emphasized when Black Moon is in Libra. Here we remember Lilith's role as handmaid to Inanna, said to bring men in from the fields for the sacred marriage rites. A wedding photographer with Black Moon in Libra considers her role to be a sacred function as she bears witness to and documents this high moment of union.

As the first wife of Adam, Lilith shows another side of herself in a marriage that lacks mutual respect and empowerment. Lilith deserted Adam, and Eve was created. Was the theme of equality resolved? Eve was blamed for the Fall. The relationship between Adam, Eve and Lilith has been explored in art ever since. There are so many ways to interpret the relationship dynamics in that original Garden of Eden, where duality was born. Two sides of the feminine dwell in the psyches of women and men, as we struggle to understand ourselves, our desires and what we want from each other.

Perhaps this duality of the feminine reflects the Libra placement between Virgo and Scorpio. Virgo tends toward the "virginal," the purity of organic self-referential experience that is brought into relationship. Scorpio tips The Scales the other way. Once a decision has been made or a relationship entered, Scorpio dives in all the way.

The polarity and union of female and male is the basic dance of life. Black Moon in Libra experiences relationships on many levels. She places things in proper relationship to each other—people, ideas, experiences—comparing and contrasting in order "to understand what all the combinations are about, how everything connects," says one native. The Laika people in Peru talk of *ayni*, which means to be in right relationship with ourselves and others, with Earth and community, in every interaction in our lives. Then we are in balance with Heaven and Earth. The Aries clarity of self in relation to others is a key to balance. Relationship with Self is another.

Lilith Black Moon is related to the Goddess of Wisdom who weighs your heart with the Feather of Truth, like Egyptian Maat, Grecian Athena, and our modern blind-folded Justice. Lilith sees right through the blind-fold with her inner eye, the third eye that transcends duality. Overshadowed by this image of the Mediatrix, these people can have a special talent for linking opposite poles, i.e., Nelson Mandela. A Harvard-trained M.D. practices primarily homeopathy and Anthroposophic medicine. He has worked in holistic clinics as well as state hospitals and prisons, keeping a foot in both worlds. A best-selling author explores esoteric concepts in adventure novels, time-traveling between the past and the present to see how history moves in circles and between people. While in the Peace Corps, a corporate business consultant for a World Bank subsidiary began training with tribal shamans. At a certain point the contrast between his two worlds required a critical choice. His work has become the education, awakening and empowerment of a shared dream of healing the way we live in relation with the Earth.

If you have Black Moon in Libra your ability to dialogue, to mediate, negotiate and foster connections is a central aspect of your life path. This doesn't mean the path is easy. Poet Robert Frost spoke of the road "less traveled by, and that has made all the difference." [see notable Black Moons in Libra]. It seems sometimes that significant people or partners disappoint or disappear. Black Moon in Libra, with its Aries opposition, pushes you back to the experience of yourself. You can't accommodate others beyond a certain point. Certain closest relationships are plumbed to the depths in the internal conversation of I and Thou. Your mystical side may call you to union with the Divine Other, even in the context of a relationship when two become one. As Rumi wrote: "Out beyond ideas of wrongdoing and rightdoing, there is a field. I will meet you there."

Aquarius

You flash like lightning over the highlands,
You throw your firebrands across the earth,
Your deafening command, whistling like the South Wind,
splits apart great mountains.
—Wolkstein and Kramer, *Inanna, Queen of Heaven and Earth*

Black Moon in Aquarius charges up a storm of electricity, amping up the chakras, meridian grids and nervous system with high, solar-powered winds. This is a placement of the Initiatrix, the goddess who pours living waters of life from her electrically-charged cloud banks, she who invented acupuncture, also shock therapy. The objective intelligence of the Black Moon can be at its best in this sign of the higher mind and broad perspective.

Capacity for multi-dimensional communication may lead to dialogue with flora, fauna, elementals or archetypes, even extra-terrestrials and other forms of subliminal information circuitry. A Capricorn Black Moon pairing will augment the connection to Nature, while a Pisces true Black Moon will open out to the universal aetheric sea, pulling in remote frequencies.

A strong sense of community weaving comes from perceiving the nesting of the microcosm in the macrocosm. Black Moon in Aquarius persons often have energy fields that buzz and give off lots of photons, as the molecules in their bodies frequently go into an excited state. Electric appliances may react to them. They may be extraordinarily sensitive to weather, particularly atmospheric fields as well as group dynamics. They may be half out of their bodies a good deal of the time and need ways to ground such high-wired electricity. Body therapies that address subtle channels are useful. The opposition from Leo suggests benefits from creative expression and looking for ways to channel their unique frequency in meaningful social ways. Harry Potter author J.K. Rowling is an example of this, as well as influential modern philosopher Jacques Derrida. For the highly individual sign of Aquarius, especially with any eccentricities indicated by the Black Moon, the community context of life is an issue of importance, even when society is engaged from an isolated distance. Modern technology can be helpful here, in proper dosages.

A woman healer alters time and energy through her presence. Her body is often in disorder because she is embodying so much extra-terrestrial and archetypal energy, as she and others perceive it. Her physical system periodically limits her computer time. Demonstrating her Black Moon in active opposition with planets in Leo, an idealized image of this unusual blond woman is portrayed in mythic landscapes quite popular in Japan, where the Sun Goddess is very deep in the national soul.

A writer channels information from other star systems in a framework that addresses the multi-dimensional and pan-historic physicality of human experience. An architect designs and builds community centers that enhance social interaction. He incorporates historic and modern design and materials in a unique way. Though a prominent member involved with his community, he is not at ease in intimate relationships and is not entirely comfortable in his body.

If you have Black Moon in Aquarius, your range of perception is as broad as the universe, encompassing particular concentric circles and your own inimitable nodes of connectivity. You may feel like an alien in a strange land, especially when surrounded by others who do not tune into such a rarified bandwidth, so obvious to you. Consider yourself a cosmic lightning rod and seek appropriate grounding by working with your body, as well as outlets for the powerful insights that come through. You find friendships in far-flung places and may have a penchant for mathematics, theoretical utopias or progressive social vision.

For the subtle energy interactions of your Black Moon, lifelong collegiality and friendly companionship can be at least as important as deep intimacy. Erotic, passionate love, whether sexualized or not, is a bigger challenge for the Aquarian need for space and individual freedom.

In Water Signs

> *When She danced on the waters, and the wind was Her horn,*
> *The lady laughed, and everything was born. . . .*
> —*Songs for the Old Religion*, words adapted by Aidan

Lilith in this fluid element is elusive and naturally psychic, flowing through unusual and delicate tonalities of feeling. She channels awareness through the fluids of the body, the blatant blood, sweat and tears as well as the more subtle lymphatic, cerebro-spinal and interstitial fluids. Such exquisite sensitivity can lead to agonies and ecstasies beyond the experience of most other persons.

Cancer

> *Lilith offered Eve the fruit of life. "Eat me."*—M. Kelley Hunter

Cancer Black Moon is the image of the Dark Mother in all her sweet, fearsome, and awesome glory, Mother of the world, its protector and sustainer. She plants seeds of Mystery in our hearts and souls. She is the life stream. She dips into the ever-swirling matrix of potentiality to bring forth what is needed, encouraging us to drink deeply. Ever-hungry and thirsty for what only she can give us, we all depend on her support for our very lives. Thus she offers the fruit of life and is herself the fruit. And when our time is up, she swallows us holy.

Family is important to Cancer Black Moon. Inherited patterns run deep; the spiritual ancestry is remembered. One may feel exiled from the blood family and seek to find one's true soul family. There are a special few with whom these persons can truly bond. Personal demons appear in life, sometimes in the form of those closest to them. Mother and mothering, or lack thereof, are key experiences. The personal mother is often overshadowed by the Great Mother. As a biological parent or not, one needs to find a way to give to others. There is a deep capacity to nourish the soul of others, but an aversion to dependency or attachments. Heads will roll, their own included, with the swipe of the sword of the protector, cutting away bonds that offer no soul growth. Here is the mother bird that pushes her baby bird out of the nest when it's ready to fly. The opposition of Capricorn faces Black Moon Cancer with its personal "dark night of the soul," exaggerating its insecurities, and calls for the development of a calm inner strength and emotional maturity that knows when and how to "draw the line" in co-dependent situations.

With Black Moon in the Moon's sign, there is a deep desire to resolve the apparent contradiction between the subjective personal mode of Cancer and the impersonal universality of Black Moon. "Intensification of spirituality

Weaver of Dreams

transforms old stuff into affirmation," says one Black Moon Cancer woman. "It takes watching your consciousness where it magnetizes back to old patterns and creates emotions that are undesirable, and then rechanneling it to cultivate a new level of consciousness." This process takes time, like churning butter, yet one can feel the change in a lightening of the emotional body. The Crab can become crabby and moody, rather than flowing with the inner tides of the Black Moon. With Black Moon leaning into Gemini any moods can be Mercurial quick-silver. Black Moon in Gemini's gift of intelligence can enhance the ability to articulate Cancer feeling impressions for excellent writing or teaching skills. A Leo flavor gives a creative flair and expression of emotional fullness, as heard in the music of Chopin, who had this combination.

A Black Moon in Cancer cook owned a natural food restaurant on a small island. The Garden of Luscious Licks was creatively decorated with inspirational writings, angels, an altar, photos and materials about ecological, holistic living, thereby feeding the soul as well as the body. Another woman feels like she is always giving to others, starting with her emotionally clinging mother. She has a hard time feeling her own needs, and has turned to the practice of psychotherapy for her own healing and that of others. The Great Mother bestows her riches and blessings in a multitude of ways.

If you have Black Moon in Cancer you *need*. Just what you need and from whom is the question. To deeply investigate your true needs from early unmet needs that generated dependencies are an inevitable exercise in detaching from past reactive patterns and improving relationships. The sensitivity of Cancer is most often felt in the stomach. Observe your eating habits. Be aware of what you eat, and when and how you nurture yourself. Revel in the act of eating. It is a sacrament every time, as we receive from the bounty of Nature, our Mother.

Composer Frederik Chopin, tragic actress Sharon Tate and priest-philosopher Pierre Teilhard de Chardin are found in this Black Moon category.

Scorpio

> *Be—and yet know the great void where all thing began,*
> *The infinite source of your own most intense vibration,*
> *so that, this once, you may give it your perfect assent.*—Rainer Maria Rilke

Black Moon in this deepest and darkest of water signs is the Ecstatic Goddess, with a "passionate freedom so strong, that it liberates us from the guilt, shame, and inhibitions of the past," writes a Scorpio Lilith therapist. "It is the juice we need to reclaim our bodies," the subtle life-giving fluids and secretions of the blood mysteries and endocrine system.

Black Moon Scorpios are deep sea divers. Some do it literally, like Jacques Yves Cousteau. Some delve into mysteries of the psyche and spirit, like John Bradshaw and Marion Woodman. This placement is a psychic and emotional workout, as surges and currents of the Scorpionic waters pull one down ever deeper. Scorpio Black Moon fights rational thought forms tooth and nail, as it passionately seeks to escape all entrapments where the vitality of life is squelched. This is not a conscious decision, and it may not be welcome when the outer supports are stripped away in circumstances that can appear as betrayals. The passionate soul seeks to regenerate and redeem something larger than self. Seeking the source of the Great Mystery, Black Moon in Scorpio people are psychic investigators in some form or other. Trust is an issue, as they are forced to trust subtle instincts that fly in the face of rational common sense. They rely more on uncommon sense.

Highly charismatic, these people can have quite an effect on others, in both public and personal relationships. The sexual charge can be strong, sometimes rechanneled, as they seek connection at a soul level and are well aware of the tantric nature and life-altering implications of sexual interchange. The Taurus sign opposition, especially when occupied, gives strong sexual desires and full-bodied expression. Lilith in Scorpio processes the ripple effects of the intimate transference of fluids and the resulting psycho-spiritual implications, always transformational.

Transformation is a key word for this Black Moon position, capable of extraordinary wizardry and magic, as it surfaces from deep inner processing that goes far deeper than most. They are concerned with motivations, having themselves come out of the dark into the light. They deal with the power of desire, for the greater good or not. The best strategy with this power is to surrender it to the God/dess, sometimes a life or death decision.

Lilith in Scorpio is often compelled to investigate the psycho-dynamics of other persons or involve themselves in behind-the-scenes power plays, as did Madeleine Albright and Hugo Chavez. Especially if Lilith leans into Libra, Black Moon seeks active dialogue about inner processes and can be a good listener, excellent qualities in a psycho-therapist or ambassador. A Sagittarian influence takes the intense focus of Scorpio and aims it like a Zen archer, pulling the bow back with heightened intention to reach the most desired target.

Jungian psychologist Marion Woodman has such a passion for life she literally danced her way out of a wheel chair. Maya Angelou transformed her traumatic early life to become a spiritual leader and author of inspirational works. One man with this placement is a leading brain researcher; his passion for life gives him amazing vitality that strikes one immediately in his presence. Another writer left a career as an army officer to train people in psychic skills to promote peace. Though these examples are all writers, their true teachings go beyond words to the immediate experience they convey.

If you have Black Moon in Scorpio, your will is a tool to hone, as is the power of desire. These are also the greatest challenges. Unconscious desires—for money, sex, position, power or even a simple parking space in front of the coffee shop—can run your life until you reach down inside to seek their source and recognize your inner motivations. Then you can consciously rearrange the subatomic energy fields to manifest what you want. Eventually, in this lifetime or the next, you will let go of all desires except the one truest one for spiritual illumination.

You likely have uncanny and penetrating insight into the psycho-spiritual processes of others as well as your own. You are your own guinea pig for such investigations. Death is a character in your life. *Meet Joe Black* actor Brad Pitt has mean Black Moon in Scorpio. There can be a fascination with the "dark" side, with the Scorpio tendency to go where angels fear to tread. Black Moon Lilith leaves no stone unturned, no matter what creepy, crawly things she finds. Through your own process of shedding ego layers like a snake skin, you find ways to empower others. This brings out the thrilling best of Black Moon in Scorpio.

Pisces

> *Far away, antennas hear antennas*
> *and the empty distances transmit...*
> *Pure readiness. Oh unheard starry music!*—Rainer Maria Rilke

Part sea-serpent, this Melusine mermaid sings songs from the ancient seas. She holds a conch-shell as her horn of plenty, calling the sound of creation from the center of the center of the galaxy, echoing everywhere in the cosmic sea. Black Moon in Pisces flows along this invisible continuum, which can be disorienting, as one is often channeling frequencies from remote sources, constantly experiencing altered space-time reality on a subliminal, superconscious level. A large part of one's attention is elsewhere. Lilith in Pisces may feel exiled on Earth, that she operates better "out there." This position shape-shifts quite easily, according to circumstances and the fluctuations of cosmic wave lengths.

Operating primarily on Dream Time, Pisces Black Moon can have a long, prehistoric sense of time. Ideals can be high and elusive. Artistic expression of some kind, be it color, music, images, can be a means for channeling the spiritual-magical energy, as can mysticism and meditation practice, or a blend of such contemplative activities. A heightened sensitivity to life's suffering can offer itself in compassionate service or sacrifice, or be lured into addictive escapism, a siren song to disaster.

The body may be extra sensitive, the aetheric body expanded and more permeable to influence from many octaves. The Virgo opposition calls for attention to the physical instrument, highly saturated with aetheric waves, as it may develop particular symptomatic aversions to certain activities that are not fulfilling on a spiritual level, requiring a change of life style, profession or environment. Relationships to people as well as to nature have a strong spiritual dimension, which might be quite subtle and mysterious, expressed in some unique form. A second placement in Aries instills a need for psychic clarity that has a fiery edge in asserting its pristine territory. An Aquarian quality more often seeks group involvement and kindred spirits.

"Beauty is the essence of life," says one woman. "I want everything around me to be made into beauty, to feel the love, the communion you can get from beauty." For some years, "forced into operating on an earthly level," she became a business rep for a creative advertising director. That ended abruptly. She loves her flower garden, feeling an intimate communion with creatures and plants—and with water, very healing for her.

A spa director, massage therapist and singer-songwriter developed Graves disease, with notable physical changes that pushed her into exploring another level of holistic and shamanic healing, including color and sound therapies. With an M.A in Counseling Psychology, she periodically finds herself working in the corporate world, feeling exiled from her spiritual purpose, while seeking a soul-satisfying synthesis of her many talents. She is a medical intuitive.

Public persons who show the potentials of this placement include Mikhail Gorbachev and Yitzhak Rabin, who shared a Nobel Peace Prize, as well as musicians Gloria Estefan, Judy Garland, Tina Turner, Robert Plant and Herbie Hancock.

If you have Black Moon in Pisces, levels of dream and reality merge. Immersion in baths or, even better, in the sea is very healing. From the water you drink to the tears that you shed, watch your own personal tides. With Virgo as the opposite sign, careful attention to the body may not only be a good thing, but a necessary one. You must learn to close up holes that allow unhealthy or even dangerous energies to come in. An orderly environment can be relaxing, making it easier to deal with daily duties. Clean on the physical level supports psychic clarity. *Caritas*, unconditional neighborly love, is your gift to the world in whatever form it takes.

Chapter Three

Famous Black Moon People

This section is a listing by sign of notable persons with prominent Black Moon positions. Witnessing these Lilith-influenced lives as much or as little as we can from the outside offers clues about how the Black Moon shows up in real life. Many of these persons exude a charismatic, mysterious or provocative presence. They are publicly known because of their profound influence on their times, for better or worse. Some have made an indelible impact. So many of these brief profiles deserve a fuller treatment, beyond the scope of our discussion here. There is plenty of fodder for further research.

Those with Black Moon on the Ascendant often show the face or features of Lilith in their physicality, walking onto the stage of life overshadowed by this Dark Goddess. The persona carries a strong impact that invites a strong reaction or response, sometimes of extreme attraction or revulsion, or both.

Those with a Sun-Black Moon conjunction will find some Lilith theme expressed along their life path. When infused by Lilith's subtle hues, the Sun glows with an ineffable "something" through the life force of these individuals, creating a "hallmark" effect. When the life purpose is colored by Black Moon, which is neither a purpose nor a reason nor any such definitive path, one is led on a labyrinthine journey. Lilith leaves her mark. Such persons fascinate us and leave a special trace behind them, not without sufferance. Esoteric lore speaks of the "Sun behind the Sun," a deeper Selfhood that is interlaced with the cosmic web. Black Moon may darken the

Sun's obvious purpose, the better to unveil and reveal its hidden dimension. Like the alchemists that seek to turn lead to gold, life's bitter moments offer some sort of redemption and release.

The Moon-Black Moon can be one of the most challenging and potentially rewarding positions because the impersonal insistence of the Black Moon overshadows the personal considerations of the Moon. The Black Moon-Moon dynamic, so naturally intertwined in their astronomical dance, is particularly and personally impersonal, and inevitably challenging. The personal Moon in relation to the impersonal Black Moon suggests a complex relationship with one's mother, as if she is overshadowed by the Great Mother in some way. Family patterns, relationships or the ancestral lineage contains complexities, including unconscious dependencies, that intimately impact one's life. The growth of the soul ultimately takes precedence over personal comfort. True communion and one's spiritual lineage may be felt outside the safety of the family hearth, or even when fostered within it. There is a deep capacity for giving, receiving and/or taking.

Depending on the nature and closeness of the aspect between these two lunar energies, their combined effect is felt from early on, even before birth. Progress the Moon backwards or forwards to the exact aspect and note the timing. (The progressed Moon moves one degree a month. Go backwards or forwards from the Moon to the Black Moon. The number of degrees is applied to the months before or after birth.) With such pre-verbal experiences, the meaning is not easily apparent. One can only be assured that there is a deep impersonal calling that seeks fertile ground somewhere in life. House placement gives a clue.

One woman with this conjunction described her early feeling that her mother took from her rather than gave, creating a mutual clinging from unmet needs—on both sides. She grew up not knowing how to nurture herself and became anorexic, denying herself basic sustenance. "My nurturing qualities (Moon) are at times taken away, sucked out of me like Lilith sucked blood with her vampirish qualities." She has come to realize her mother was not fully able to nurture herself, either. This woman now works in equine-assisted psychotherapy.

Note: The mean Black Moon is generally used as the leading Lilith in this chapter, although the true placement is noted when different. Ascendant examples are based on birth times being correct. Refer to the data table at the back of this book. Except in regard to the Ascendant, the time of day does not affect mean Black Moon sign placements, so birth data without a firm time can be used. True Black Moon positions move more quickly, but are confirmed as listed. The AstroDataBank program, much appreciated for its extensive listings and Rodden rating system, was an indispensable tool for chart verification. *Merci beaucoup* to Astrotheme.com, a site which allows Black Moon searches, a blessed aid to this research. All data is listed at the end of this book. Charts are from SolarFire with the mean Black Moon position and an arrow indicating the true Black Moon posiiton. When time is unknown, natural houses are used, with 0 Aries on the first house cusp, and based on noon local time.

Black Moon in Aries

Note the originality in the lives of these people who pushed their own edges and struck a new note in the collective consciousness. Black Moon in Aries people must take risks, whether with relish or reluctance, sometimes showing up ahead of their times, living with courageous flair and sometimes a bit of fire-breathing.

Black Moon-Ascendant in Aries

John Lennon (see Black Moon Beatles section)

Gene Simmons (true Black Moon in Pisces)

"The Demon" of the punk-rock band KISS puts on a Lilith look for his act: fire-breathing, blood-spitting, tongue-wagging with ritualistic face paint.

Willie Brandt (true Black Moon in Aries, mean Black Moon in Pisces)

This resistance fighter became a leading post WWII statesman, leading Germany to restored world power. He was awarded the 1971 Nobel Peace Prize for his initiatives in negotiating non-aggression treaties with the USSR and Poland and between West and East Germany.

Black Moon-Sun in Aries

Loretta Lynn (true Black Moon in Taurus)

Iconic American country-western singer, Lynn's autobiography, *Coal Miner's Daughter*, made into an Oscar-winning Hollywood movie, documented her life from a poverty-stricken upbringing in rural Kentucky to super-stardom. She developed a signature voice, "a mature fusion of twang, grit, energy and libido," as her website describes it, delivered with guts and humor. The first woman named "Entertainer of the Year" by the Country Music Association, Lynn was a major force in bringing a female perspective into country music. She sang and wrote about women's issues such as birth control, marital abuse, divorce, war, "life as it was lived—in the kitchen and in the bedroom—by millions of working-class women everyday." By age fifty, she had sixteen #1 songs and fifty-two top-10 hits. a songbook for Lilith in Aries: "You Ain't Woman Enough to Take My Man," "Don't Come Home A'Drinkin'," "Fist City," "I Wanna Be Free," "Rated X," "The Pill." Several of her more controversial songs were initially banned. Married to Oliver "Doolittle" Lynn at age fourteen, she had six children. The marriage lasted forty-eight years, through thick and thin, until her husband's death. Her twin daughters, Patsy and Peggy, became country singers, as did Loretta's sister, Crystal Gayle. Her 2002 autobiographical sequel is entitled *Still Woman Enough*.

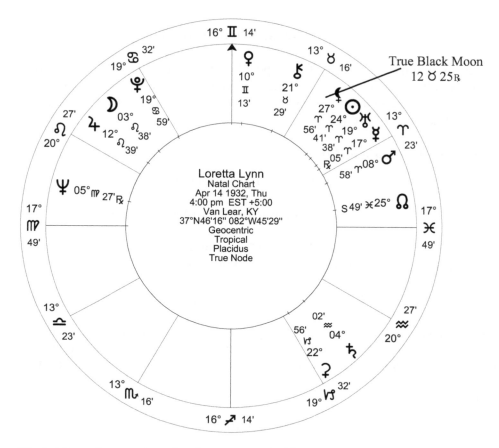

Loretta Lynn
Natal Chart
Apr 14 1932, Thu
4:00 pm EST +5:00
Van Lear, KY
37°N46'16" 082°W45'29"
Geocentric
Tropical
Placidus
True Node

True Black Moon
12 ♉ 25ℝ

Her mean Black Moon-Sun conjunction is accompanied by Mercury and radical Uranus, this generational planet in exact square with Pluto in Cancer, showing that she was a transformational pioneer for the times. True Black Moon is well into Taurus in close square to Jupiter and wide square with her Moon in Leo, beautifully reflecting her life path, her show-womanship and her heart.

Francisco de Goya (true Black Moon in Taurus)

This great Spanish master artist had Aries Black Moon conjunct his Sun, as well as Venus, Mercury and Mars. This multiple conjunction was challenged by a square with Moon and Neptune in sensitive Cancer. He painted for Spanish royalty and also chronicled the horrors of the military history of his era. Later in life, illness, both physical and mental, led to the "Black Paintings," depicting the dark side of the times and his fantasy. Astrologers should know his "Saturn Devouring His Son" from 1819. Beginning in the style of the Old Masters, Goya came close to pure Expressionism in later works, a modern artist ahead of his time.

Prospero Lambertini, Pope Benoit XIV

One of the most scholarly of the Popes, he was an unusually free thinker for his time. As Pope from 1740-1758, he authorized publications about the revolutionary heliocentric world view and wrote works himself to align Catholic doctrine with the new science of his time.

Alec Guinness (true Black Moon in either Aries or Taurus)

This Oscar and Tony award-winning actor has Black Moon conjunct the Sun, and the likely dawn birth time would put it right on his Ascendant. As an illegitimate child with a very unsettled and confusing childhood, he may well have the Black Moon rising. In his extensive career he played all types, from the Pope to Hitler, kings and ghosts. Sir Alec was knighted in 1959. He won a 1957 Academy Award as Best Actor for *The Bridge On the River Kwai* and the 1977 Best Supporting Actor Oscar for playing Obi-Wan Kenobi in *Star Wars IV*. Sir Alec received an Academy Honorary Award for lifetime achievement in 1980 and became a Companion of Honor in 1994, a British and Commonwealth Order recognizing outstanding achievements.

Omar Shariff (true Black Moon in Taurus)

This exotic-looking actor, best known for his movie role of Dr. Zhivago, was also a highly regarded professional bridge player, writing books and columns in that field. The Black Moon is very close to his Sun, Uranus and Mercury (retrograde). Raised Catholic, he became Islamic in order to marry.

Gary Oldman

An excellent example of Black Moon Aries, which is closely flanked by both Sun and Moon in the tenth house, this British actor has created an astonishingly original array of movie characters, inventing a unique voice or accent for each. Often there is a Lilithian dark edge to his roles. He has played such characters as musician Sid Vicious of the Sex Pistols to Dracula, Beethoven, a pimp, terrorist, corrupt and sadistic officials and Sirius Black in the Harry Potter movies. He wrote, directed and produced the largely autobiographical and highly praised film, *Nil by Mouth*. Abandoned by his own father at a young age, he later found himself the single father of two young sons.

John Bobbitt

This man's life is a warning to men who dishonor Lilith! In a notorious American scandal, his wife Lorena sliced off half his penis in a fit of revenge after he drunkenly raped her.

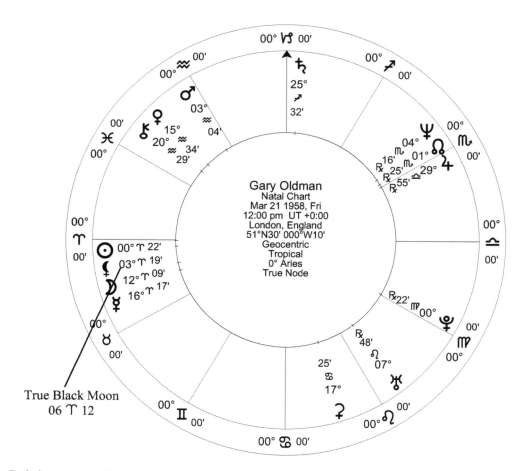

Gary Oldman
Natal Chart
Mar 21 1958, Fri
12:00 pm UT +0:00
London, England
51°N30' 000°W10'
Geocentric
Tropical
0° Aries
True Node

True Black Moon
06 ♈ 12

Driving away, she threw the body part into a field and then called police. The penis was located and reattached in a nine-hour surgery. Bobbitt recovered; the couple divorced. She testified in court to his years of infidelity, abuse, a forced abortion and selfish sex, leaving her unsatisfied. He was acquitted of abuse. She was also acquitted, her crime of passion judged to have occurred in a fit of insanity. Bobbitt's Black Moon in Aries is conjunct his Sun-Saturn in Aries. The night of the incident, June 23, 1993, mean Black Moon was on the degree of Bobbitt's Chiron in Pisces, activating the archetype of the Wounded Healer. Bobbitt went on to form a musical band, The Severed Parts, in an unsuccessful effort to pay his medical and legal bills. In that year of 1993, the Uranus-Neptune conjunction at 19-21 Capricorn was square Bobbitt's true Black Moon. This case was one of the first to legitimize issues of marital rape and domestic violence.

Black Moon-Moon in Aries

Rabindranath Tagore

"I came out on the chariot of the first gleam of light, and pursued my voyage through the wilderness of worlds leaving my track on many a star and planet."—from "Journey Home" in *Gitanjali*

This beloved and prolific Bengali writer and educator won the Nobel Prize for Literature in 1913. The first Asian to win the award, he broke new ground by writing in his local language and with a psychological realism new to Indian literature; both are suggested by the Moon-Black Moon conjunction. His mystical poems (in *Gitanjali* and *The Gardener)* are best known in the West, but his extensive work includes novels, plays, essays and songs. Politically active, he promoted ideals of trans-nationalism and "unity consciousness." Dismayed about the state of education, he founded Santiniketan, a highly successful alternative school. If the early morning time is correct, Black Moon was also close to his Ascendant.

Nella Jones

This accurate clairvoyant started to use her unusual gifts later in life, working as a psychic for the British police, experiencing both fame and skepticism. At the end of an Aries stellium of Mercury, Uranus and Mars, her Moon and true Black Moon join at 29 Aries. The Sabian symbol for this degree is relevant to her psychic knowledge: "A celestial choir has arisen to sing cosmic harmonies." Mean Black Moon is a few minutes into practical Taurus, heading toward her Sun.

Are you one or do you know one?

Many persons born between November 1975 and March 1976 have Black Moon conjunct Jupiter with Mercury and/or Chiron. These natives are on a mission streaked with originality, "where no one has gone before."

More Black Moons in Aries

Madonna, Holly Hunter, Patrick Stewart, Michael Jackson, Albert Einstein, Mohandas Gandhi, Bruce Lee, Elizabeth Taylor, Marcel Marceau, Jean Paul Sartre, Rudolph Steiner, Bhagavan Shri Osho Rajneesh, Henri Matisse, Pele, Placido Domingo, Andrea Bocelli, Marc Chagall, Joan Baez, Le Corbusier, Christian Barnard, Edgar Degas, Jennifer Capriatti, Georgia O'Keefe, William Bouguereau, Maria Malibran

Black Moon in Taurus

In this sign Black Moon feels the Earth energy specifically and deeply. Nature calls, through a tangible, intimate connection of human nature and the cosmos. The esoteric motto for Taurus is, "I see and when the eye is opened, all is light." The Bull's Eye is the star Aldebaran, a red supergiant star that brings awareness of the bare simple fact of existence, of death in life and life in death. Loss transforms to calm acceptance. Fully engaged with the stuff of life, embedded in physicality and highly appreciative of its sensuality, Black Moon Taurus at best sees through material substance to the life force that generates it.

Black Moon-Ascendant in Taurus

Georges Bernanos

> *Little things seem nothing, but they give peace, like those meadow flowers which individually seem odorless but all together perfume the air.*—from *The Diary of a Country Priest*

A two-degree conjunction of both Black Moons and his Ascendant gave this man a very intense, Lilith sort of look. An important and original Roman Catholic writer of the twentieth century, Bernanos wrote such prize-winning works as *La Joie* (*Joy*) and his masterpiece, *The Diary of a Country Priest*, made into an award-winning movie. *Under the Sun of Satan* (a great Lilith title) was also made into a film. He was a fanatic monarchist, supported by a Moon-Pluto conjunction in Gemini opposed by Jupiter in Sagittarius.

Black Moon-Sun in Taurus

Catherine II, "The Great"

This powerful and tyrannical Empress of Russia reigned for thirty-four years, expanding the borders and political power of Russia. She took the throne in a coup from her incompetent and reportedly impotent husband. A true Taurus patron of the arts, literature and education, she established the Russian Ballet, as well as a school for girls, and corresponded with leading thinkers in Europe. A typically highly-sexed Taurus, she was generous to her many lovers, which she took even into advanced age. The Russian word for money, *babki*, means "old women" and refers to her picture on old ruble bills, appropriate for Taurus.

Empress Eugenie

The last French empress, wife of Napoleon III, was born in Spain with a long list of royal titles. Known for her intelligence, beauty, charm and extravagance, she was a fashion leader and advi-

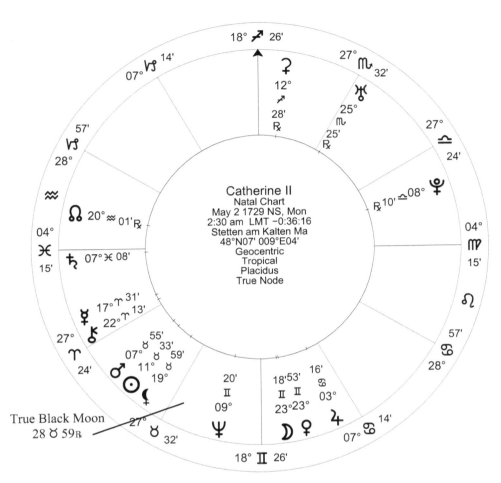

sor to her husband. After the Franco-Prussian War, she lived in exile for the last forty years of her life. In a famous incident at a Bournemouth, England spa, a path of hundreds of tea candles were set to light her way to the sea, a continued tradition. Asteroid 45, discovered in 1857, was named Eugenia in her honor, the first asteroid to be named for a living person, and one of the first asteroids found to have a moon. The tiny orbiting body, discovered in 1998, was named Petit-Prince for the famous book, *Le Petit Prince* (*The Little Prince*) by Antoine de St. Exupery. That book was written for Eugenie's son, the Prince Imperial.

David Hume (true Black Moon in Aries)

"Where men are the most sure and arrogant, they are commonly the most mistaken."—from *An Enquiry Concerning the Principles of Morals*

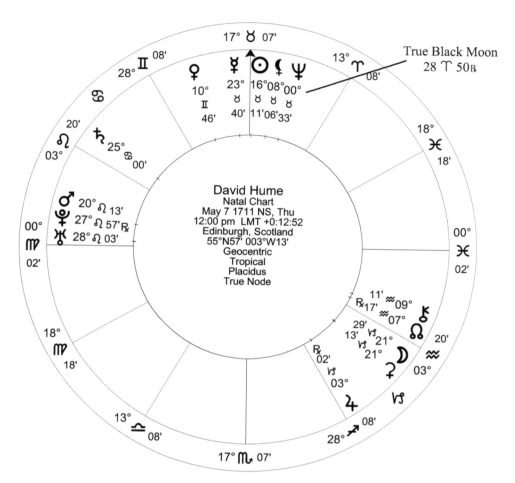

Philosopher, historian, economist and leader of the Scottish enlightenment during the Uranus-Pluto conjunction of his era, Hume is considered one of the most important persons in the history of Western philosophy, the first to fully articulate a modern naturalistic approach. He wrote *A Treatise on Human Nature* at age twenty-six, a very intense work that caused bouts of depression. His six-volume *History of Great Britain* was a best seller and stood unsurpassed for more than sixty years. The Black Moons on either side of the Aries-Taurus cusp speak both to his pioneering work and the naturalistic theme.

Many 1906 Taureans have Sun conjunct Black Moon. Examples include:
Roberto Rossellini (May 8): This leading Italian film director and producer was a major influence in neo-realist cinema. He was known as well for his scandalous affairs and marriage to Ingrid Bergman.

Mary Astor (May 3): A starlet of the 1920s, she won an Oscar for Best Supporting Actress in 1941 for her role in *The Great Lie* and played the irresistible villainess in the classic, *The Maltese Falcon*, with Humphrey Bogart. Her career faltered at times due to personal problems, including scandals, drinking, health issues and family problems. Her first husband died in a plane crash. She experienced three divorces, custody battles and a suicide attempt. Her parents, German immigrants who lived a lavish life off her earnings, sued her for support. She wrote two best sellers about her life and made television appearances well into the 1960s. Mean Black Moon was widely conjunct Sun in Taurus. True Black Moon in the first degree of Gemini joined Venus, Mars. and Jupiter while also square the Moon, giving her sparkle, wit, precociousness and adding another layer to her family problems.

Kurt Godel (April 28): Considered one of the most important scientists of the twentieth century, with interests in philosophy and physics, Godel revolutionized the logical foundations of mathematics. A refuge from Nazi conscription, he emigrated to the U.S. in 1940. A friend of Einstein, he later shared the first Einstein Award in 1951, was awarded the National Medal of Science in 1975, and received honorary doctorates from many prestigious universities, such as Yale and Harvard. Although mean Black Moon was widely conjunct the Taurus Sun, it was much closer to Venus and Mars in Taurus. The true position is telling to his work, conjunct Jupiter in Gemini. His intelligence was multi-disciplinary. Also, because born in Brn, Moravia in the early twentieth century, his citizenship changed from Austrian to Czech to German and, finally, American.

Black Moon-Moon in Taurus

As the Moon is exalted in Taurus, its conjunction with the Black Moon is especially interesting.

Rene Descartes

"Cogito, ergo sum. I think therefore I am."

With mean and true Black Moons closely conjunct ruling Venus and exalted Moon, this man had a good grasp of reality! His famous philosophical statement has endured through the centuries, extolling the fact that what we think defines our experience of reality. His Aries Sun was accompanied by Uranus, Jupiter and Pluto for a pioneering spirit that invented analytic geometry.

Are you one or do you know one?

Many 1941 Taureans have a close Black Moon conjunction with Sun and Venus. This group includes:

Ann-Margret: The career of this glamorous Golden Globe award-winning actress, singer and dancer has spanned decades.

Ryan O'Neal: This famed veteran actor is the father of Academy Award winning actress Tatum O'Neal.

Judith Blegen: A beautiful leading American operatic soprano, she has the conjunction very close, along with the Moon.

More Black Moons in Taurus

Julia Roberts, Martha Stewart, Vincent Van Gogh, Agnetha Faltskog, Lucy Lawless, Mary Cassatt, Henry David Thoreau, Ella Fitzgerald, Stevie Wonder, Orson Welles, Amelia Earhardt, Gabriel Byrne, Willie Scott, Abraham Lincoln, Edgar Allen Poe, Paul Verlaine, Richard Branson

Black Moon in Gemini

Gemini Black Moon speaks with many tongues, a trickster who is part news reporter, part salesperson and part spy. As disparate and seemingly distracted as this sign can be in its curiosity for life, Lilith in Gemini essentially seeks the vertical alignment of higher and lower minds in order to find unity in the diversity. These busy bees make a rare, medicinal and occasionally divinatory, sweet, sometimes bitter, honey from the unique pollens they collect.

Black Moon-Ascendant in Gemini

Christine Keeler

This call girl/model was the key figure in the Profumo affair, a sex scandal that toppled Great Britain's Conservative government of Harold Macmillan in 1963 and made sex a hot political potato. Her autobiography became the subject of songs and films. In 1987 she appeared on a promotion video for Bryan Ferry's "Kiss and Tell" hit single.

Gene Wilder

Nominated for Academy Awards as an actor and a writer, Wilder is most known for his work with Mel Brooks, as co-star in comedies with Richard Pryor, and as Willy Wonka in the original 1971 movie, *Willy Wonka and the Chocolate Factory*. Married for five years to comedienne Gilda Radner, who died of cancer, he continues to promote awareness and treatment of that disease. When we add in his Sun conjunction, we can call him a "double" Black Moon.

Black Moon-Sun in Gemini

Paul McCartney (see Black Moon Beatles section)

Rafael Nadal (true Black Moon on Taurus cusp)

This top Spanish tennis player and Olympic gold medalist, known as "King of Clay," has a Sun-Black Moon-Chiron stellium. Gemini is a tennis type of sign with its ball bouncing back and forth across the net and its multiple playing surfaces. Nadal's powerful play has also earned him the nickname, "The Matador," a nuance of the true Black Moon in Taurus.

George Herbert Walker Bush

Forty-first President of the U.S., from an oil-rich family, "Bush the First" has Black Moon in Gemini on his Midheaven with Gemini Sun in the tenth house. The youngest Navy combat aviator in WWII, he was awarded a Distinguished Flying Cross. Heading into politics, he was elected to the U.S. House of Representatives, served, as Ambassador to the United Nations, Chairman of the Republican National Committee, Chief of the U.S. Liaison Office in China, Director of the CIA and Vice President under Ronald Reagan. As President he headed the U.N. coalition in the Gulf War, "Desert Storm." His famous campaign promise, "Read my lips. No new taxes," was often cited as (Gemini) duplicity.

Sir Laurens Van Der Post

". . . the loneliness creeping into the heart of modern man because he no longer sought the answers of life with the totality of his being."—from *The Heart of the Hunter*

Born with Sun in Sagittarius *opposite* a Black Moon-Pluto conjunction, this famous Afrikaner of Dutch-German descent was a farmer and storyteller, conservationist, journalist, explorer and writer. Early in his life he was disturbed by the prejudice of the educational system that continued the racial divide in Africa. His life had several dramatic chapters, set between South Africa, England and the war years. As a prisoner of the Japanese in Java, he helped organize an educational program. His books and documentaries on the Kalahari Bushmen are well-known. At times a political advisor to the British government, he also came to know the Bloomsbury writers, Carl Jung and Prince Charles, among other luminaries. He was godfather to Prince Harry. He died at the age of 90, whispering "die sterre" (the stars). After his death, moral questions and controversy about the truthfulness in his writings arose illustrating the Sagittarius-Gemini duality of truth vs. fudging facts for a good story, a blend of fact and fiction into myth. Inevitably the Sun-Black Moon opposition will shed light on the dark side and throw shadows across the Sun, especially in the dualistic sign Gemini and even more so with Pluto close by.

Black Moon-Moon in Gemini

Hugh Jackman

This Aussie actor and producer is extraordinarily versatile. In musical theater, he wowed London as Curly in a stage version of the classic American musical *Oklahoma!* and New York as composer Peter Allen in *The Boy from Oz*. He has hosted the Tony Awards. Best-known as a film actor, his most famous role was Wolverine in *X-Men*, with a set of knives for a hand—very Black Moon in the "handy" sign of Gemini! In *The Prestige*, a dark movie of magic and illusion, Jackman played two characters competing in stage magic against a pair of twins. In *The Fountain*, an esoteric movie challenging perception, he plays three life-times of a man searching for immortality and eternal love. In the Gemini myth, one twin is mortal, one immortal. Jackman will leave a legacy of great work to immortalize him.

Helen Keller

Left blind, deaf and dumb after a childhood illness, this amazing woman became an inspiration as deaf educator and world ambassador. Her Black Moon in Gemini square Moon in Pisces contributed to her access to alternate forms of communication, as did her gifted teacher Ann Sullivan.

Friedrich Nietzsche (true Black Moon in Taurus)

> *"All truth is simple . . . is that not doubly a lie? "*—from *Maxims and Arrow*
> *"Against boredom even the gods struggle in vain"*—from *The Antichrist*

This famous philosopher and classical philologist was a radical thinker, whose influence came in highs and lows. He raised profound questions that few of his contemporaries could keep up with. His opposition of Gemini mean Black Moon with Moon in Sagittarius is conjunct the nodal axis with the Black Moon on the South Node. Thus he was born with a Lilith legacy. This opposition makes a T-square with a Venus-Chiron conjunction in Virgo. All of the feminine indicators—Moon, Venus, and Black Moon—are in this challenging T-square pattern, suggesting an active feminine side within himself, leading him to question personal, social, and religious values. Women played a key role in his life. When he was a young boy, his father and brother died, leaving him in a household with his mother, grandmother, two aunts, and a younger sister, with whom he had a complex relationship. His adult relationships with women were difficult. The inherent quality of Gemini results in the discrepancy between ideas in his published vs. unpublished works and still unresolved opinions about them.

The true Black Moon in Taurus supported his pragmatic, sensate philosophy which pitted vital life experience against religious after-life idealism and led to the existential movement. Yet this

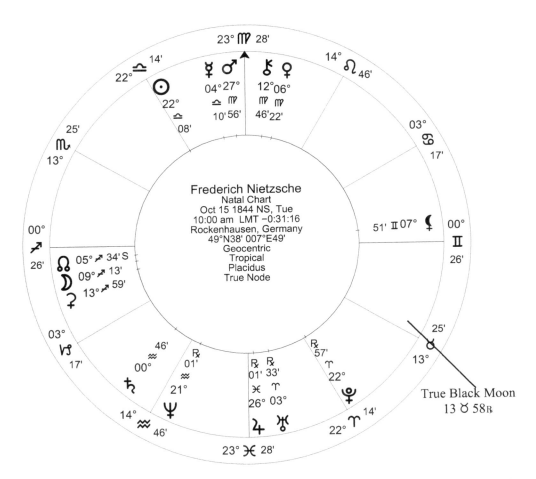

Frederich Nietzsche
Natal Chart
Oct 15 1844 NS, Tue
10:00 am LMT −0:31:16
Rockenhausen, Germany
49°N38' 007°E49'
Geocentric
Tropical
Placidus
True Node

True Black Moon
13 ♉ 58ᴿ

earthy Black Moon opposed his Scorpio Sun, forming a T-square with Neptune in Aquarius that led to his recognition of the "will to power" and his advocacy of creative artitsry. The Sun-Black Moon-Neptune pattern contributed to his genius and his sensitivity, but also his sense of being misunderstood, loosening his grip on reality and leading to his eventual psychotic breakdown. His influence continues. "That Nietzsche was able to write so prolifically and profoundly for years, while remaining in a condition of ill-health and often intense physical pain, is a testament to his spectacular mental capacities and willpower," records the Stanford Encyclopedia of Philosophy. "Lesser people under the same physical pressures might not have had the inclination to pick up a pen, let alone think and record thoughts which—created in the midst of striving for healthy self-overcoming—would have the power to influence an entire century."

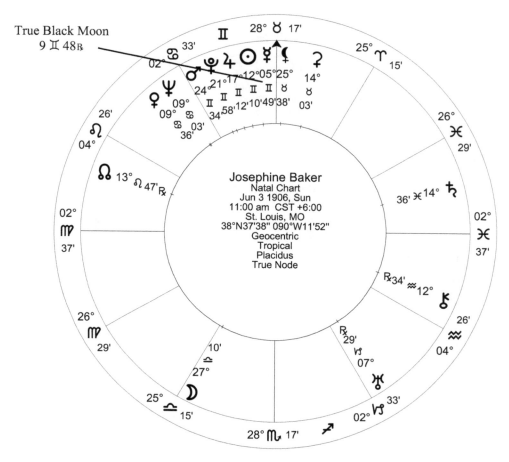

True Black Moon
9 ♊ 48℞

Josephine Baker (true Black Moon in Gemini, mean Black Moon in Taurus)

Driven by dark forces I didn't recognize, I improvised, crazed by the music. . . . Each time I leaped I seemed to touch the sky and when I regained earth it seemed to be mine alone."—from *Josephine,* her autobiography

From an impoverished childhood in the highly racist United States of the early 20th century, married twice by the age of 14, this independent-minded and astonishingly brave and original Lilith woman took every opportunity she was offered, becoming the "first Black superstar" of stage, screen and recording. Her life accomplishments include honors and medals as a member of the French resistance and her work as a civil rights activist in the U.S. Ahead of her time, she adopted 12 foster children from around the world to create her "Rainbow Tribe," living in a re-stored castle in France. Her menagerie of animals included cheetah, pig, chimpanzee, goat, snake, birds, cats and dogs.

Starting with the Black vaudeville circuit in the midwest, Freda McDonald Baker found her way to New York City in 1921 in the midst of the Harlem Renaissance. First a comedy chorus girl in the popular touring show, "Shuffle Along," she was soon invited to be a dancer in "La Revue Negre" in Lilith-loving Paris. She was an overnight sensation after her famous debut wearing only a pink flamingo feather. Her uninhibited and provocative dance movements contained humor, infectious energy and an exotic beauty that inspired other artists and epitomized her age, a fusion of the Black artistic renaissance, jazz and modernism. She enjoyed fantastic success in Europe and South America. Signature pieces included the banana dance, for which she originally wore only a skirt of bananas, suggestive of the phallic excitement generated by her erotic shimmy.

Still too provocative for the United States of the 1950s, she was condemned in the McCarthy era and exiled for nine years, returning to speak at the 1963 March on Washington, followed by successful performances to mixed black and white audiences only, on her insistence. She had become a French citizen through the third of her five marriages. Her lack of financial management necessitated continued work, yet she was eventually evicted from her castle with her 12 adopted children and offered sanctuary in Monaco by Princess Grace. Her 1973 farewell performances at Carnegie Hall in New York, in London and in Paris were sold out. She died of a stroke in the middle of the Paris run. As a sub-lieutenant of the French air force, she was given a 21-gun-salute at her funeral in Paris.

It was true Black Moon on her Gemini stellium of Sun, Mercury, Jupiter, Pluto and Mars, that gave steamy entertainer Josephin Baker her star of destiny. The mean Black Moon in Taurus lent the fearless sensuality and earthy charisma of this iconic "Black Venus," "Creole Goddess." Her signature song, in true Gemini style, was "J'ai Deux Amours, mon pays et Paris," I Have Two Loves, my country and Paris."

Are you one or do you know one?

Many Geminis born in the following years have a Sun-Black Moon conjunction: 1924, 1933, 1942, 1951, 1977, 1986, 1995 and 2004. 1942 and 1977 featured the addition of Jupiter. Geminis born in 2004 share the influence of Venus retrograde with the conjunction. Venus is the esoteric ruler of this sign, and thus brings a rare emotional perception and rich nuances into the life path.

More Black Moons in Gemini

Shirley Chisholm, Barbara Streisand, Harrison Ford, Charlie Chaplin, Ramana Maharshi, George H. Bush, Frank Sinatra, Celia Cruz, John Huston, Korla Pandit, Sean O'Casey, Garrison Keillor, Jan Vermeer, Golda Meir, Basia, Lillie Langtry, Michael Faraday

Black Moon in Cancer

Those with this position have a special relationship with the Great Mother. In Buddhism, the Mother of all the Buddhas is said to be she who takes us beyond the beyond, giving birth to consciousness and bliss. This Mother is the cornucopia of all that nourishes life, especially the life of the spirit. This is not a comfortable placement, but brings many gifts to the one that shares her or his own.

Black Moon-Ascendant in Cancer

Marilyn Horne

This mezzo-soprano operatic star is largely considered to be one of the greatest singers in the history of opera. Conjunct her Cancer Ascendant, Black Moon is also opposite her Capricorn Sun. She established the Marilyn Horne Foundation to help preserve the art of vocal recitals, demonstrating the Cancer-Capricorn tendency to protect tradition.

Sharon Tate (true Black Moon in Leo)

This up-and-coming young actress was the wife of Roman Polanski. She was murdered when eight months pregnant by the Helter Skelter gang of Charles Manson in the notorious multiple murder. Swiss Astrologer Claude Weiss gives a good lecture on the various Black Moon aspects of this horrific event. Tate's mean Black Moon was conjunct Jupiter just behind her Ascendant in Cancer.

True Black Moon was leaning into the first degree of Leo, conjunction Pluto and opposition a twenty-minute Mercury-Sun conjunction. Her mother became a victims' rights advocate, influential in amending criminal law in California to allow crime victims and their relatives to make victim impact statements at parole hearings. I always thought it odd that Polanski's eerie movie, *Rosemary's Baby,* came out just a year before this tragedy.

Kim Wilde

This award-winning recording artist turned gardener has become a popular cult figure in the U.K. Her first recording in 1981, "Kids in America" made her a star in the wider European market. She has opened for Michael Jackson and David Bowie and spent a year singing in the musical, *Tommy*. While pregnant she studied gardening and was picked up by BBC for a gardening show. She published *Gardening with Children* in 2005. The mean Black Moon and Mars are conjunct her Ascendant and trine her Sun-Moon in Scorpio.

Black Moon-Sun in Cancer

Merv Griffin

U.S. television personality Merv Griffin hosted one of the first major talk shows. A man of his nation, Griffin was a double Cancer with an exact Sun-Pluto conjunction rising and on the July 4 United States Sun. Black Moon joined the conjunction, all opposite a Full Moon in Capricorn exactly conjunct Jupiter. This is a strong signature in any case, with the Black Moon adding an extra essence of public appeal and deep personal meaning. There is a story that on his eighteenth birthday, he heard a "miracle voice" that told him he would never again be a private person, a challenging message for a double Cancer.

Jacques DeLors

Another man of his country and beyond, Jacques DeLors has Black Moon and the Sun conjunct the Moon. Savvy economist and politician, DeLors helped revive the French economy in the 1980s, served in the European Parliament, and went on to lead an unprecedented two terms as president of the EU Commission. The DeLors Commission forged today's Euro economy. He also chaired the UNESCO Commission on Education in the 21st Century, overseeing that committee's report entitled, "Learning: The Treasure Within." If the suggested birth time of 12:30 p.m. is correct, the triple conjunction tops his chart at the Midheaven.

Roald Amundsen (true Black Moon in Leo)

The famous adventurer who discovered the South Pole was also the first to sail through the Northwest Passage and possibly the first to fly over the North Pole (even though Richard E. Byrd is credited for the latter). Demonstrating the Cancer tendency for safety and protection, even in such extreme life-and-death journeys, Amundsen said: "Defeat is certain for him who has neglected to take the necessary precautions. . . . Victory rewards him who has everything in order—luck, people call it." With mean Black Moon conjunct Sun, he also had true Black Moon exactly conjunct Jupiter in Leo and on his Ascendant if the 5:00 a.m. birth time is correct.

Black Moon-Moon in Cancer

Arthur Rimbaud

A French visionary poet, Arthur Rimbaud wrote some of the most exquisite verse in French literature. His Black Moon in Cancer was square his Moon in early Libra, double trouble in "normal" living, suggesting a challenge to emotional security and relationships. A brilliant student, he had a difficult mother and absent father and often ran away. His meeting with the Symbolist

True Black Moon
4 ♌ 52ℝ

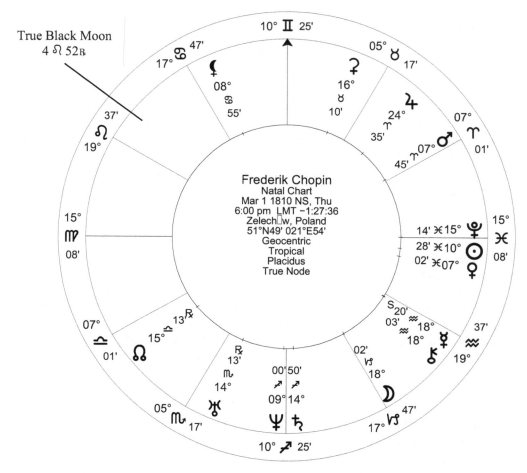

poet Paul Verlaine turned into a tempestuous love affair, liberally laced with absinthe and hashish, that fueled his poetic vision. However, the relationship turned ugly, and Rimbaud had Verlaine arrested. He spent the next several years writing his best works, innovating in free verse. At age twenty-one he stopped writing to go off adventuring. He died at age thirty-seven.

Frederik Chopin (true Black Moon in Leo)

Chopin's sensitive Black Moon in Cancer was opposite his Moon in Capricorn and part of a grand trine in water signs with Uranus in Scorpio and his Sun, Venus and Pluto in Pisces. Doesn't his music sound like that, with its heightened emotional flow and moods? But there is more. He also had a Lilith fire triangle, with true Black Moon in Leo, Mars in Aries and Neptune in Sagittarius. Mars was square the mean Black Moon. The Lilith-infused water and fire created a highly-charged poetic emotionality, wonderful for art, challenging for relationships. He had a

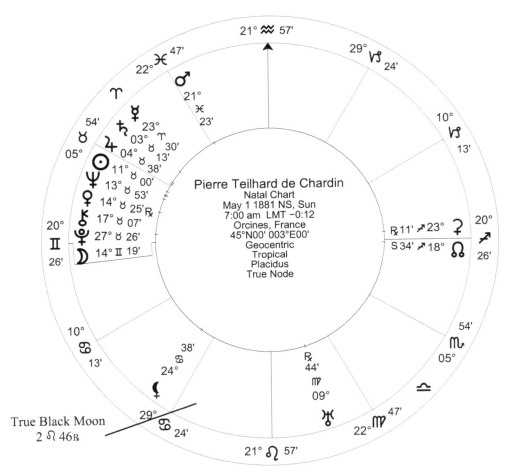

Pierre Teilhard de Chardin
Natal Chart
May 1 1881 NS, Sun
7:00 am LMT −0:12
Orcines, France
45°N00' 003°E00'
Geocentric
Tropical
Placidus
True Node

True Black Moon
2 ♌ 46ʀ

passionate nine-year affair with a free-spirited female writer, George Sand. Her Cancer Sun was right on his Black Moon. Chopin died at age thirty-nine of tuberculosis, leaving a rich and extensive legacy of some of the most moving piano music ever written. His tomb at Pere La Chaise cemetery in Paris is topped by a sculpture of a sorrowing muse and constantly decorated with flowers, candles and other mementos of appreciation.

Pierre Teilhard de Chardin (true Black Moon in Leo)

Blessed be you, universal matter, immeasurable time, boundless ether, triple abyss of stars and atoms and generations: you who . . . reveal to us the dimensions of God—.from "Hymn to Matter"

Scientist, mystic, and Jesuit priest, Teilhard de Chardin was a visionary of the twentieth century.

His life path was motivated by a passionate sense that spirit and matter are one, evolving to merge in the light of revelation in spirit. From early in life his ecstatic experiences of nature informed his studies in physics, geology, paleontology, and chemistry. With his Black Moon in Cancer square Mercury in Aries, he uniquely articulated the true soul of science with a sense of a living, loving cosmos. Appreciation for the cosmic feminine was an intimate experience for this Christian mystic, silenced by the Roman Catholic Church for his unorthodox views on evolution and consciousness. The true Black Moon in a square to his Jupiter-Sun may indicate the unease of religious authorities in the face of his love of *anima mundi*.

Are you one or do you know one?

The New Moon on July 4, 1951 was conjunct the Black Moon as well as Mars and Uranus. Fireworks are an appropriate birthday celebration! Cancers born from July 11 to 20, 1934 have Pluto and Mercury retrograde joining their Black Moon and Sun, especially potent the first two days. Deep thinkers, these folks.

More Black Moons in Cancer

Barbara Hand Clow, Giorgio Armani, Anjelica Huston, Gustave Courbet, Kenneth Branagh, M.C. Escher, Robert Schumann, Henry VIII of England, Katherine Hepburn, Rembrandt, Raphael, Piet Mondrian, Geoffrey Rush, Pir Vilayat Khan, Antonio Banderas, Catherine Zeta-Jones, Jennifer Lopez, Audrey Tautou, Edward Norton, George Harrison, Sting, Janis Joplin

Black Moon in Leo

Ruled by the Sun on the exoteric, esoteric and soul-directed levels, Lilith in Leo seeks to expose the soul's fiery creative power either greedily controlled by or divinely channeled through ego identity. The glory of true queenship or kingship is granted to those who lead and are led by love shining in the clear mirror of the heart.

Black Moon-Ascendant in Leo

Marilyn Monroe

Playboy's #1 sex star of the twentieth century, Marilyn Monroe's indefinable magic and unique Lilithian charisma translated almost tangibly in film and photographs. Her sex symbol image was belied by the seriousness of her commitment to the craft of acting. She was adept at comedy and drama, reaching deep inside herself to pull out nuanced characterizations. She was intent on finding herself and used acting as a means to do that. Neptune joined the Black Moon on her Ascendant to add more glamour and dream woman illusion. That conjunction was opposite Moon

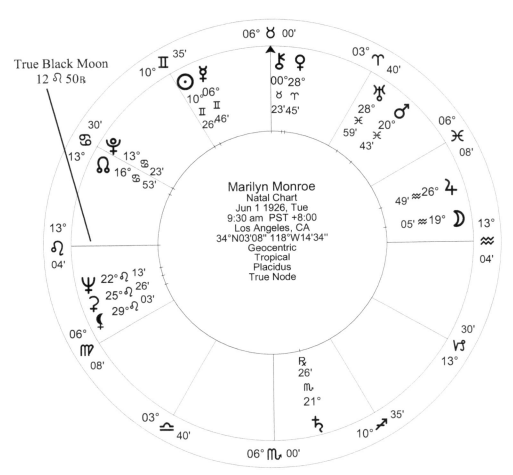

True Black Moon
12 ♌ 50ʀ

Marilyn Monroe
Natal Chart
Jun 1 1926, Tue
9:30 am PST +8:00
Los Angeles, CA
34°N03'08" 118°W14'34"
Geocentric
Tropical
Placidus
True Node

and Jupiter in Aquarius. Her mother was mentally unstable. Her liaisons with men of power were well-known. She died under mysterious, veiled circumstances in 1962. Her continuing iconic status is unprecedented.

Pablo Picasso

A main founder of the Cubist art movement, Picasso delighted in disrupting the aesthetics of classical art as he brought into expression the chthonic influences underlying the twentieth century. With mean and true Black Moon on his Ascendant, his life and art truly portray the face of Black Moon Lilith.

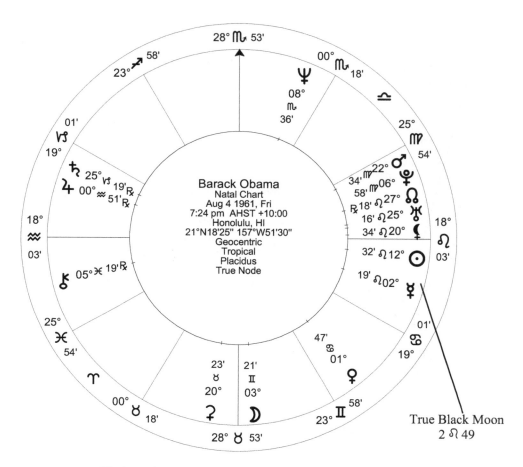

Christopher Reeve (true Black Moon in Virgo)

Superman super star, this American actor and director demonstrated the super-hero in film and real life. He became a regal Leo role model to the world in the face of a tragic accident that left him paraplegic, dramatizing the extremes of Lilith from his wheelchair. He continued to work, directing and acting and to advocate for the disabled until his death. He regained a modicum of movement many years after the injury, the first recorded such case, for which he credited his work with the mind-body connection. We can also credit his true Black Moon in Virgo.

Black Moon-Sun in Leo

Barack Obama

With his Sun at the midpoint between his true and mean Black Moons, the 44th President of the United States is a Lilith leader. The mean Black Moon is conjunct Uranus and the North Node as well. He is completely unexpected on the world stage, a bi-racial and multi-national man with a white American mother and African father, early years spent in Indonesia, later raised by his grandparents in Hawaii. From the first he has demonstrated unorthodox and skillful leadership style.

The true Black Moon in the same degree as Mercury accounts for his extraordinary intelligence and raises the Gemini Moon to a further level of eloquence. This conjunction is opposite Jupiter in Aquarius, enlarging his social vision and global awareness, suggesting the legal education and community development work that started his career. Jupiter retrograde suggests an inherent wisdom that draws from some cultural heritage or wisdom tradition from the past, but that needs to be brought up to date for the modern times. Obama's election and presidency has relied heavily on internet communications technology and the social networking revolution of the early 21st century. The T-square to Neptune from the Leo and Aquarius polarity suggests his finger on the public pulse and his ability to articulate compelling mottos ("Change we can believe in") as well as the shrewd perception and political savvy necessary to accomplish his agenda. The Neptunian spiritual ideals and sensitivity connected to Black Moon go beyond religious ideologies, or else he can be blinded or deceived by the very ideals he proclaims. He aspires to government transparency. Let's see how far he can go with that.

The Inauguration chart for the first January 20 swearing in (remember we had two, a Mercury retrograde reassurance) shows the mean Black Moon in the same degree as Pluto at the early cusp of Capricorn, opposite Obama's natal Venus in Cancer. He inaugurates a new era. The financial crisis is of primary concern, a Venus matter. On another level, this Lilith-Venus opposition points to the key role of the women in his life, wife, daughters and mother-in-law, the family in the White House, Though we know she was born on January 17, 1964, we don't know the time of Michelle Obama's Chicago birth. The Moon changed from Aquarius to Pisces that day. What do you think? If born around noon, the Moon has moved into Pisces to join Venus and square Lilith in early Sagittarius. There is no doubt, just from seeing the mean Black Moon in Barack Obama's 7th house, that he has a Lilith First Lady.

Alfred Hitchcock (true Black Moon in Virgo)

Born on a day with Sun and Black Moon closely conjunct, with Venus nearby, this ultra-famous film director brought the aura of Lilithian mystery to the silver screen in dozens of classic movie mysteries. His lugubrious look made him immediately recognizable and suggests the veracity of a pre-dawn birth that puts Black Moon in the first house, maybe even closer to his Ascendant than the 3:15 a.m. time gives. He was an intensely shy and retiring person. Perhaps his true Black Moon in Virgo accounts for that. Conjunct a ruling Mercury and squaring Uranus in Sagittarius, we can certainly say that true Black Moon contributed to his ground-breaking vision and technical skill.

Rose Kennedy

This well-educated socialite, ambassador's wife and matriarch of the famous American political family gave birth to nine children and lived to the age of 104. Two sons became U.S. senators and one a president. She lost one son in WWII and two sons by assassination (John and Robert). One daughter died in a plane crash, as did her grandson, John Kennedy, Jr. This was in addition to living with the numerous affairs and public scandals of her husband. Her close Sun-Mercury conjunction was joined by Black Moon Lilith, all opposite Jupiter in Aquarius.

Black Moon-Moon in Leo

Queen Latifa

With both Black Moon positions conjunction her Moon, of course this iconic, Grammy Award-winning rap musician would call herself "Queen" and name her debut album *All Hail the Queen*! Tall and full-figured, this well-known film and television actress has a strong, dramatic persona. She wrote a 1997 book on self-esteem entitled, *From the Heart of a Queen*. This conjunction claims extra attention, being the focus of a T-square with Saturn/Mars in Taurus imposing from one side and Jupiter in Scorpio from the other.

Are you one or do you know one?

Many 1943 Leos, including those mentioned below share a dramatic stellium of Sun-Jupiter-Pluto-Black Moon-Mercury-North Node:

Mick Jagger (July 26): This lead singer of the legendary Rolling Stones still rocks on after 40 years. His theatrical, Leonine presence is an intricate part of the act, along with long-lasting songs and sound. The true Black Moon in Cancer sextile a strong Moon in Taurus suggests the importance of his Black Moon family connections (see Black Moon-Moon in Pisces for daughter Elizabeth).

Robert DeNiro (August 17): The strong Lilith aura of this well-known American actor shines in such movies as *Taxi Driver, The Scent of a Woman* and *The Godfather*. A life-long New Yorker, he is a driving force behind the revitalization of the Tribeca district, a once-neglected warehouse district, and, after 9/11, the Tribeca Film Festival, that attracts a world-wide audience. Thus he demonstrates the heart-of-gold leadership of Leo.

Caril Ann Fugate (July 30): Living the dark side at a young age, Fugate has become the stuff of myth and movies. At age thirteen, she became the accomplice of serial killer Charles Starkweather, participating in the deaths her three family members before heading out on a murder spree. She was the youngest female to be on trial for first-degree murder. A model prisoner, she was paroled in 1976 and now lives a quiet (normal?) life.

Are you one or do you know one?

Other notable 1943 natives, including those listed below, do not have the Sun in Leo, but still demonstrate the power of the Black Moon in multiple conjunction. Are you in this 1943 super Leo group? How do you express it?

Nikki Giovanni (June 7): *"If I could come back as anything—I'd be a bird, first, but definitely the command key is my second choice."* This resilient Gemini woman became a major voice in the Black Arts movement of the 1960s, connecting literature and politics, winning such awards as the NAACP Image Award for Literature and the Langston Hughes Award. Her chart shows a Moon-Pluto-true Black Moon conjunction at 5 Leo. Her poem, "Ego Tripping (there may be a reason why)" (such a Leo title!) gives powerful expression to the goddess, to Earth and to the Leonine queenship of Lilith in Leo:

> *I designed a pyramid so tough that a star*
> *That only glows every one hundred years falls*
> *Into the center giving divine perfect light*
> *I am bad. . . .*

Lech Walesa (September 29): Once an imprisoned union activist, this man became leader of the Solidarity union, and then president of Poland from 1990-1995. He won the 1983 Nobel Peace Prize. Born on a Libra New Moon with a Neptune conjunction, he has a strong stellium of Pluto, Black Moon, North Node and Jupiter in the twelfth house. He is said to be devoted to the Black Madonna of Czestochowa, nationally honored as Queen and Protector of Poland since 1656.

More Black Moons in Leo

Jan Vermeer, Princess Diana, Queen Elizabeth II, Pierce Brosnan, Michael J. Fox, Jane Austen, James Joyce, Patrick Swayze, Eddie Murphy, Johannes Kepler, Isaac Newton, Angela Lansbury, Sri Aurobindo, Sri Mukhtananda, Carlos Castenada, Ho Chi Minh, Victor Hugo, Michelangelo, Charles De Gaule, Brigitte Bardot, Julio Inglesias, Jim Morrison, Joni Mitchell, George Clooney, John F. Kennedy, Marquis de Sade

Black Moon in Virgo

The Black Moon in this earth sign can hone its finest skills and/or exhibit its most neurotic traits. Lilith seeks to fully experience the nuanced sensuality inherent in Virgo; when suppressed, problems ensue and health suffers. Attention to detail can be exquisite and/or painful. The High

Priestess performs her rituals to honor nature in psycho-somatic experience connecting cosmos and psyche in mundane life.

Black Moon-Ascendant in Virgo

Charles Dickens

This great English novelist portrayed the dark side of poverty and the working class poor in his many enduring novels. He wrote from his own experience, as his father went to debtor's prison, and he went to work on the streets at age thirteen. A moody man, he had ten children with a wife he disliked and eventually abandoned to pursue an affair with a young actress. The sensual side of Virgo cannot be dismissed in the presence of the Black Moon. He had true and mean on his Ascendant square Jupiter in Gemini.

k.d. lang (true Black Moon on Leo cusp)

This sultry Canadian songstress of many styles has a twelfth house cluster. Mean Black Moon is just minutes into Virgo, within a degree of her Moon and close to Pluto, all just behind the Ascendant and opposite Chiron in Pisces. Appropriately, her Grammy award-winner was called "Shadowland." The true Black Moon is back in the last Leo, between Uranus and the North Node, accounting for more of her eclecticism and eccentricity. Her Black Moon territory is the Leo-Virgo cusp, the cusp of the Sphinx, creature with lion body and the head of the goddess.

Seiko Matsuda

This Japanese super star made a record twenty-four consecutive #1 hit singles. One of the most influential artists in Japanese pop history, she has received more media attention than anyone else in the country. Though trying for success abroad, part of her appeal is specifically cultural. She embodies the Japanese "burikko," a woman who acts like a cutesy girl to be attractive, even into her thirties and forties. She has Black Moon and Pluto straddling her Ascendant, sign of a cultural icon. She divorced, a rare event in Japan frowned upon when initiated by the women; yet it reinvigorated her popularity in the late 1990s.

Black Moon-Sun in Virgo

Ken Kesey

This Lilith man headlines a rare group of 1935 natives born with a complex Lilith stellium in Virgo consisting of retrograde Venus, Neptune, and Sun. This legendary actor, novelist and leader of the Merry Pranksters became an iconic figure of the 1960s. He authored *One Flew Over the Cuckoo's Nest* and *Sometimes a Great Notion,* both made into movies. While at Stanford University in 1959, he volunteered for a CIA-financed study on the effects of psycho-active drugs. He worked the night shift at a veterans' hospital, where he developed some of his characters. When arrested for possession of marijuana in 1966, he faked his own death and crossed the Mexican border. The Virgo stellium fromed a restless mutable grand cross opposite Saturn in Pisces, with Mars in Sagittarius and Moon-Chiron in Gemini square to both.

Karl Doenitz

This decorated German admiral led the successful U-boat campaign in World War II and succeeded Hitler as Fuhrer for less than one month at the close of the war, negotiating Germany's surrender. It was said that his tendency to micromanage (Virgo) caused some losses. He served ten years in prison for war crimes and "crimes against peace." Later he wrote two books about his life and his reputation has been reviewed and improved. His Black Moon-Sun was in a stellium with Mercury, Saturn and Venus in Virgo and opposite Jupiter in Pisces. Mars, earlier in Virgo was opposite his Moon in Pisces, both square the peaking Pluto-Neptune conjunction of his birth year. A remarkable chart made more so by the Black Moon-Sun.

Cleveland Amory

This American columnist, writer and television critic was widely known for his witty caricature sketches that appeared in Broadway playbills. He was also very involved with animal rights and co-founded the Humane Society in the U.S.

Johann Wolfgang von Goethe

Certain defects are necessary for the existence of individuality.—from *Elective Affinities*

This famous philosopher and literary giant, who brought Lilith out of obscurity in his *Faust,* left a lasting legacy in the collective consciousness and world culture. He had mean and true Black Moons closely hugging his Sun within a degree and opposition to the Full Moon in Pisces.

Black Moon-Moon in Virgo

The fourteenth Dalai Lama and exotic dancer Candy Barr share the same birth date and hour, both born with a three-degree span between the Black Moons, Moon and Neptune in Virgo, yet expressing very different dimensions of Lilith.

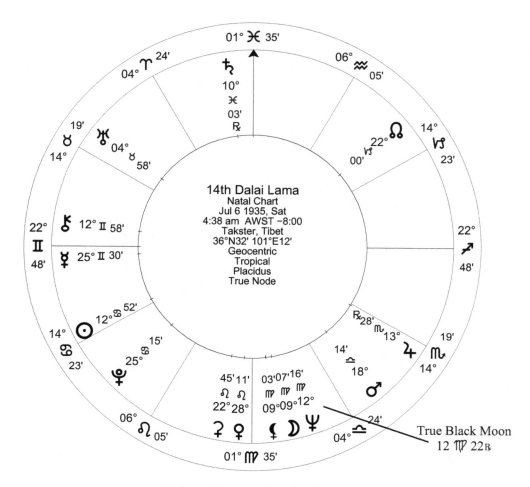

14th Dalai Lama, Tenzin Gyatso

This exiled leader of Tibet carries on a centuries-old spiritual lineage. His birth was foretold by his predecessor, said to be his previous incarnation. He was "discovered" at age two. The Dalai Lamas are believed to be manifestations of Avalokiteshvara or Chenrezig, the Bodhisattva of Compassion and patron saint of Tibet. Bodhisattvas are enlightened beings who have postponed their own nirvana and chosen to take rebirth in order to serve humanity. The Dalai Lama is the "mother" of his people in a Tibetan Buddhist sense that names Prajnaparamita ("beyond the beyond") as the Mother of all the Buddhas. Forced into exile by the Chinese invasion of Tibet, he remains the temporal and spiritual leader to his people, as well as guru to many Western Buddhists. Author of many inspiring books, he channels a deep compassion into this war-torn world and has been awarded the Nobel Peace Prize.

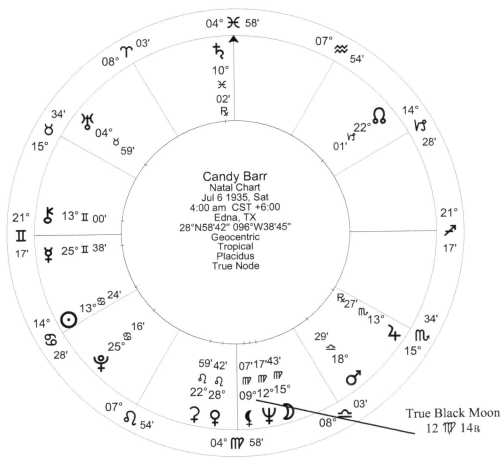

Candy Barr

A tumultuous childhood, including the death of her mother when she was nine, sexual abuse, and numerous siblings and half-siblings, led Juanita Slusher to run away and become a prostitute in her early teens. She married four times and had one child. "Loneliness is like an early frost. Let us be among the seedlings that survive," she wrote in a poem. She became famous as the Texas cowgirl stripper and burlesque dancer, Candy Barr, and is now among the inductees in the Hall of Fame of Exotic World Burlesque Museum located in California. She was said to have a perfect body. Sometimes called the first American porn star for a brief appearance in the movie, *Smart Alec*, she later served as choreographer and technical advisor to Joan Collins, who was cast as an exotic dancer in a Hollywood movie. Barr's troubles with the law included shooting of her estranged second husband and three years in prison for marijuana possession. For some years she was the girl friend of gangster Mickey Cohen and appeared in court to testify in his trial for tax evasion. After the assassination of President John Kennedy, she was questioned by the FBI about her knowledge of acquaintance Jack Ruby, who had given her a dog to begin her second career as a breeder. Ms. Slusher died December 30, 2005.

While in prison, she wrote poetry that was published in a 1972 collection called, *A Gentle Mind . . . Confused*:

> *Hate the world that strikes you down,*
> *A warped lesson quickly learned.*
> *Rebellion, a universal sound,*
> *Nobody cares, no one's concerned.*
> *Fatigued by unyielding strife,*
> *Self-pity consoles the abused,*
> *And the bludgeoning of daily life,*
> *Leaves a gentle mind . . . confused.*

Are you one or do you know one?

1935 shows up as a specialty year for Lilith Black Moon. In addition to Ken Kesey and the extraordinary astro-twins from opposite sides of the globe, we find several more examples, starting in April of that year, that suggest the special nuances of Lilith in Virgo. Of course many Virgos have a Black Moon-Sun conjunction. Are you one?

Francoise Sagan (June 21): This French novelist and playwright became famous at age eighteen upon the publication of her famous first novel, *Bonjour Tristesse*. Her Black Moon-Ascendant was joined by Neptune and opposed by Moon at the early edge of Pisces, suggesting extraordinary talent for fiction but also addictive problems, glamour and scandal. Other works with Neptune-Lilith titles were *A Certain Smile*, *Those Without Shadows* and *Those Wonderful Clouds*. True Black Moon was in Leo conjunct Venus in the twelfth house and square Jupiter in Scorpio, contributing to both artistry and self-undoing. Her extravagant life style included convictions for cocaine use and tax fraud. She lost the fortune she earned.

Joan Kennedy (September 5): A double Virgo with a complex conjunction of Sun and Black Moon with Venus (retrograde) on her Ascendant (all square Chiron in Gemini), she married into the high-profile Kennedy family with its tragedies and scandals, including the well-publicized, accidental death of her husband's mistress. She endured difficult pregnancies and several miscarriages, giving birth to three children. A 1994 biography was entitled, *Joan the Reluctant Kennedy*. After years of denial, she eventually went into alcohol recovery treatments. She and

Ted were divorced in 1982. The presence of Venus retrograde adds an internalizing emotional dimension to female experience and relationships that gives Black Moon more room to exact her depth. True Black Moon just over the Libra cusp, conjunction her chart ruler Mercury, emphasizes the impact of relationships in her life.

Julie Andrews (October 1): This famous singer and actress has mean Black Moon just behind her Virgo Ascendant, close to Neptune and strong enough to radiate through her persona. She had a remarkable four-octave range even at a young age. Becoming a professional at age twelve, she went on to star in numerous classic musicals, winning an Academy Award for her role as Mary Poppins. Interestingly, Dame Julie also has Ceres, the "nanny planet," on her Ascendant. She lost her singing voice later in life after throat surgery, but has continued to appear in movies. True Black Moon is conjunct her Libra Sun. She has been married to director Blake Edwards since 1969.

More Black Moons in Virgo

Diana Ross, David Bohm, Galileo, Alexander Graham Bell, Woody Allen, Elvis Presley, Indira Gandhi, George Roger Waters (Pink Floyd), Elizabeth Kubler-Ross, Jacques Prevert, Sheryl Crow, Matt Damon, Meg Ryan, Sigmund Freud, Tony Blair, Fidel Castro, Sai Baba, Camille Claudel, Antoine de Saint-Exupery

Black Moon in Libra

Relationship is the hallmark of this Lilith position. This includes relationship of self to other, self to Self, and self to the world. A fine balance of factors is negotiated at subtle levels. It is never more true than in this position, that what is actually going on in any relationship is impossible to judge by outside parties, or even the partners themselves. A soul contract is in play. Here, Lilith plays no favorites and makes unusual choices in life from a deep soul desire.

Black Moon-Ascendant in Libra

Venus Williams

Venus and her sister Serena have been dominant competitors in women's professional tennis for more than a decade. In addition to having chart ruler Venus (retrograde) conjunct her Sun in Gemini (did her parents follow astrology or was it destiny?), Venus has Black Moon and Pluto on her Libra Ascendant. She has won dozens of titles, including seventeen Grand Slams in singles, doubles, and mixed doubles, as well as three Olympic gold medals, two in doubles with her sister. Her relationships with her sisters have been central in her life. Her closest rival is her sister Serena, whom she has faced in numerous tournament finals, with an even record to date.

True Black Moon
12 ♎ 54ʀ

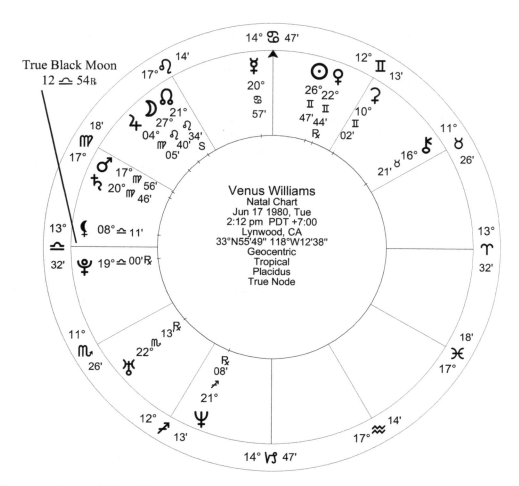

Venus Williams
Natal Chart
Jun 17 1980, Tue
2:12 pm PDT +7:00
Lynwood, CA
33°N55'49" 118°W12'38"
Geocentric
Tropical
Placidus
True Node

They are also multiple winners as doubles partners. Their older sister, Yetunde, who was personal assistant to the sisters, was shot and killed near their home practice courts.

A true Lilith legacy of her Black Moon in Libra was her highly publicized campaign for equal pay for women in tennis, which yielded results in 2007. She herself was the first woman to benefit, winning in Wimbledon that year. But she always said it was for the girls and women that would follow her.

Stylish and elegant, she also developed an interior design firm and a fashion line of shoes and clothing. She wears her own innovative and colorful sportswear while she plays. A cultural force to contend with, Venus was included in the *Ladies Home Journal* 2001 list of the thirty most powerful women in the United States.

Suzanne Valadon

Valadon was a central and colorful personage in the Montmartre art scene during "la belle epoque" of the late nineteenth century. A post-impressionist artist herself, she is best known as model and mistress of several famous artists, including Renoir, Erik Satie and Toulouse-Lautrec, and as mother of Maurice Utrillo. Her familiar face and figure are seen in numerous paintings, for instance as the dancing woman in Renoir's "Dance at Bougival." A fairly reliable dawn birth makes Valadon a double Libra with Black Moon-Ascendant conjunct Mars and North Node in her first house. Self-awareness came largely through relationship, starting with her difficult single mother. In addition to her various artistic liaisons, she tried conventional marriage with a banker for fourteen years. The love of her life was artist Andre Utter, twenty-one years younger. Her disturbed but talented son was her charge for life. On the aesthetic side of Libra, her paintings increasingly gained attention, including the admiration of Degas. As a woman, she portrayed the female body in a vital, earthy, boldly feminine, Lilith-inspired way.

Black Moon-Sun in Libra

Annie Besant (true Black Moon in Virgo)

This former president of the Theosophical Society illustrates the esoteric leanings of Lilith and the impact of significant others. Her meeting with Madame Blavatsky was a major turning point in her life. Her Uranus in Aries in the first house is opposite a complex stellium involving a close Black Moon-Sun with the North Node, Mercury and Venus (retrograde) all in a nine-degree range. She groomed the young Krishnamurti to lead the organization and was disappointed when he chose a different path and went his own way.

Josephine-Charlotte

Her Royal Highness Princess Joséphine-Charlotte, Ingeborg, Elisabeth, Marie-Josée, Marguerite, Astrid, Princess of Belgium, was born with Black Moon, Sun and Mars conjunct, became Duchess of Luxembourg in a royal marriage with Prince Jean. Thus her marriage made her life. She was born into the complex relationship dynamics of European royal families. Her mother, born Princess of Sweden and a direct descendant of Napoleon's wife Josephine, died when she was eight. The young girl studied child development and maintained an interest in the arts, childhood, family and health issues, serving as president of the Luxembourgish Youth Red Cross for many years, along with her other cultural and social responsibilities.

Josephine-Charlotte had a close Sun-Mars conjunction with both mean and true Black Moons in Libra in her twelfth house, all opposition her Moon in Aries and square Pluto. Her roy-

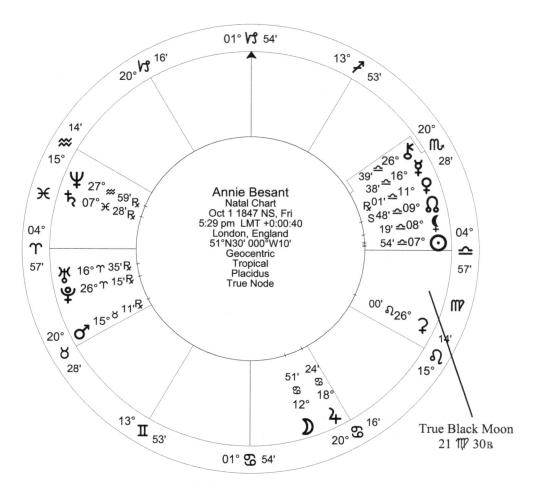

Annie Besant
Natal Chart
Oct 1 1847 NS, Fri
5:29 pm LMT +0:00:40
London, England
51°N30' 000°W10'
Geocentric
Tropical
Placidus
True Node

True Black Moon
21 ♍ 30ʀ

ally-arranged marriage to Prince Jean took place April 9, 1953, with Sun-Venus (retrograde) in Aries opposition Neptune-Saturn placed across her own Lilith-charged opposition. In addition, mean Black Moon was conjunct her Venus in Virgo. Chiron filled in the empty corner of her cardinal T-square, as did her husband's Sun-Black Moon conjunction in Capricorn. One wonders about their relationship, about any sense of personal sacrifice, if she felt constricted by her position and the intensity of the needs of others, and how fulfilling she found her service work and her cultural interests. Apparently she enjoyed gardening (Venus) and sports (Mars).

Jeanne Weber

Lilith as child killer lived through this infamous murderess. This canny nanny was accused but not initially convicted of strangling a young boy and possibly killing several other children, including her own. She fled Paris to escape an angry public that was convinced of her guilt. Later she committed more murders. Eight child murders are certain, but more are suspected. She died in a mental asylum by strangling herself "in a fit of madness." Her Libra Sun was closely conjunct true Black Moon and widely conjunct a mean Black Moon-South Node conjunction opposition Chiron-North Node in Aries. Her Sun was also square Ceres, the "nanny planet."

Paul von Hindenburg (true Black Moon in Virgo)

A German Field Marshall and war hero of World War I, Hindenburg was president of Germany during Hitler's rise to power. Influenced by his son, a key relationship for this Libra Black Moon man, he was instrumental in persuading Wilhelm II to abdicate and was pressured to form a presidential government, thereby undermining Germany's road to democracy and paving the way for the Nazi regime. Against his better judgment, he was forced to name Hitler as chancellor. The Zeppelin was named in his honor. His exact Black Moon-Sun conjunction was joined by Venus and the North Node.

Robert Frost (true Black Moon in Virgo)

Two roads diverged in a wood, and I—
I took the one less traveled by,
And that has made all the difference.—from "The Road Not Taken"

A New Hampshire farmer for many years, Frost became a premier American poet, winning the Pulitzer Prize four times. With Black Moon in Libra opposition his Aries Sun, the death of his wife impacted him strongly. From then he started attending the Bread Loaf Writers' Conference in Ripton, Vermont, an educational retreat center, spending the summers in that state every year until his death. He was named Vermont's Poet Laureate. The lines above from "The Road Not Taken," one of his most famous poems, speak of the choices we make in life. Black Moon Librans become utterly aware of the paths taken or not. His true Black Moon in Virgo was conjunct Jupiter and opposite Mercury, added to the excellence of his word craft.

Mary Pickford

Known as "America's Sweetheart," this Oscar-winning Canadian-American actress with blond curls was an international superstar of the silent screen and came to wield a great deal of power in the industry through her popularity and astute business sense. She produced films, becoming a cofounder of the United Artists film studio in 1919, of the Academy of Motion Picture Arts and Sciences in 1927 and of the Society of Independent Motion Picture Producers in 1941. She won the first Best Actress Oscar in 1929 for her first speaking role in *Coquette*, but lost public popularity when she shed her cute, "sweetheart" look. She stopped acting in 1933 but continued

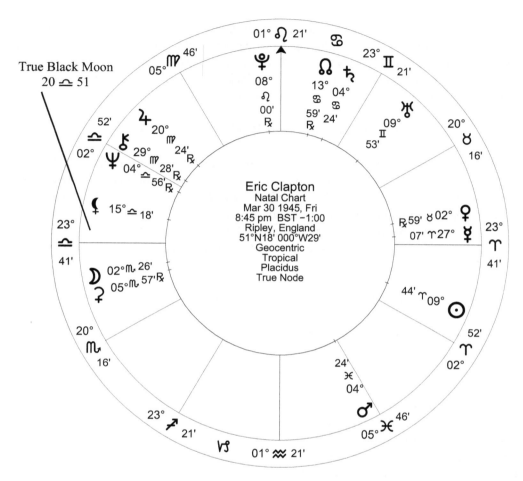

True Black Moon
20 ♎ 51

Eric Clapton
Natal Chart
Mar 30 1945, Fri
8:45 pm BST −1:00
Ripley, England
51°N18' 000°W29'
Geocentric
Tropical
Placidus
True Node

producing. Douglas Fairbanks was the love of her life. They entertained the intellectual and artistic elite in their home, Pickfair. After their divorce, she married again and adopted children with her third husband, yet Fairbanks remained her true love. Her Aries Sun opposition mean Black Moon was almost exactly square Mars in Capricorn, indicating a push-pull challenge to retain self-alacrity in relationships with men.

Eric Clapton

This Grammy Award-winning musician, considered one of the greatest guitarists of all time, has the Black Moons opposite his Aries Sun, square his Nodes and possibly conjunct his Ascendant. Clapton has made artistically fruitful collaborations and also has a rich solo career. With Derek and the Dominos, he made the classic (Lilith) song, "Layla." He was a member of the popular group Cream.

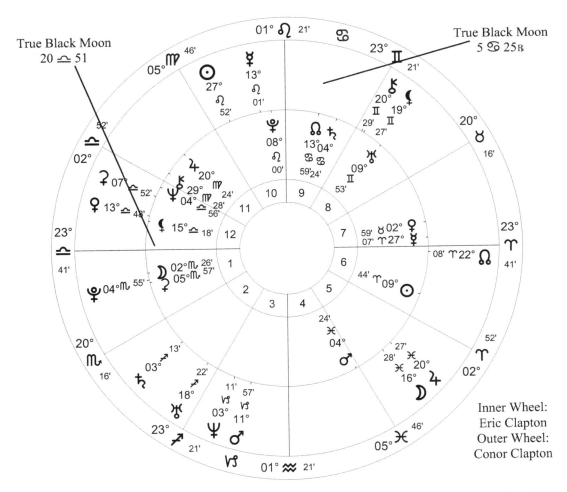

True Black Moon
20 ♎ 51

True Black Moon
5 ♋ 25 ℞

Inner Wheel:
Eric Clapton
Outer Wheel:
Conor Clapton

He was the third party in a love triangle with George Harrison and his wife, Patty Boyd, until she divorced Harrison to wed Clapton. On March 20, 1991, his four-year-old son tragically died, inspiring one of his most enduring and endearing songs, "Tears in Heaven." On that day, his Black Moon Lilith was opposed by Mercury and squared by an opposition of Chiron in Cancer and Uranus-Neptune in Capricorn. In other words, the child-killer face of Black Moon Lilith was being activated by a major T-square. Also note the exact Venus conjunct his Venus.

Chart synastry shows strong Lilith contacts between the charts of father and son. Clapton's mean Black Moon square his North Node was aligned with son Conor's true Black Moon square Venus. I believe that this activity landing in Clapton's twelfth house of karmic debts and ninth house of belief systems reinforces the likelihood that Clapton's birth time is close to correct as given here.

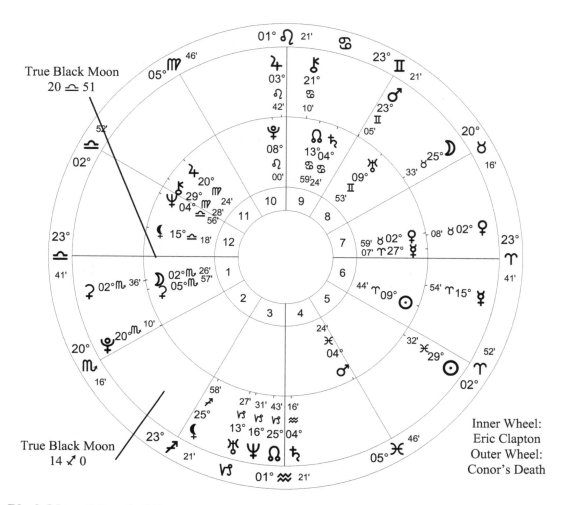

True Black Moon
20 ♎ 51

True Black Moon
14 ♐ 0

Inner Wheel:
Eric Clapton
Outer Wheel:
Conor's Death

Black Moon-Moon in Libra

Louis Vuitton

This French designer founded a luxury fashion and leather goods brand. More than a century later it retains its glamour and high-end reputation. This conjunction in his fourth house, indicating the family business, is square the 1821 Uranus-Neptune conjunction in Capricorn that evokes collective trends, and is opposite a Saturn-Jupiter conjunction in Cancer, offering popularity, recognition of quality and endurance.

Are you one or do you know one?

For those born in 1954, Black Moon Lilith joins Neptune to add her special accent to an already dynamic transpersonal T-square that formed the basic structural pattern that year: Neptune in Libra square an opposition of Uranus in Cancer and Chiron in Capricorn. The dynamic is particularly pronounced in the charts of those born that year in April with the Sun in Aries and those born in July with a Cancer Sun. Another fascinating alignment occured in 1980 as Black Moon Lilith and Pluto were conjunct the Sun of Libras born in mid October.

More Black Moons in Libra

J.R.R. Tolkien, Modest Mussorgsky, Paul Cezanne, Camille Pissarro, Nikola Tesla, Nelson Mandela, Katherine Neville, Richard Feynman, Elinor of Aquitaine, Martina Hingis, Karin Boye, Charlemagne, Kahlil Gibran, Georges Bizet, George Bernard Shaw, John, Milton, John Perkins, Ralph Waldo Emerson, Louis Althusser

Black Moon in Scorpio

Examples of Scorpio Black Moon on the Ascendant abound in the public record, as Lilith shows her seductive radiance here. Scorpio brings out the hidden intensity of any factor, and Lilith revels in the depths of this sign. There is no doubt that much goes on beneath the surface in the lives of Black Moon in Scorpio people. Passions are strong and the power of desire and sexuality is a lesson learned one way or another, sometimes on the edge of life and death.

Black Moon-Ascendant in Scorpio

Diane Keaton

This American actress of comedy and drama debuted in Broadway's *Hair* in 1968 and gained fame through her film work with Woody Allen. She won an Academy Award and Golden Globe for her title role in *Annie Hall* in 1978. She won another Golden Globe in 2004 for *Something's Gotta Give*, in which she had a rare (for her) nude shot—gutsy for a someone fifty-eight years old! With true Black Moon just one minute from her Ascendant and mean Black Moon only two degrees away, she just had to do it, especially as her Lilith Ascendant is in a T-square with her Pluto-Moon opposition in Leo-Aquarius. She is a very strong Lilith woman.

Shirley Temple Black

This beloved child actress started singing and dancing at age three. After her breakthrough movie, *Bright Eyes*, at age six, she became the largest money-making entertainer in the depression era. Her young talent was sophisticated and professionally executed. She retired from motion pictures at age 21 and later became a well-respected diplomat, serving as ambassador to Ghana, then Czechoslovakia. In 1938 she won a British libel suit against a review by Graham

Greene, who remarked on her seductive coquetry, vitality and desirable young body—his way-too-public response to her Black Moon Ascendant. True Black Moon in Scorpio is opposite her Sun-Chiron in Taurus. Mean Black Moon is in a grand water trine with Mars in Pisces and Pluto in Cancer.

Charles Lindberg

With true and mean Black Moons right on his Ascendant square Mars in Aquarius, this famous pilot courageously succeeded in the first solo transatlantic flight. He was also an inventor of medical and aircraft technologies. He was honored with the Congressional Medal of Honor, and received the 1953 Pulitzer Prize for his autobiography named after his plane, *The Spirit of St. Louis*. His reputation was tarnished by anti-Semitic comments and involvement with the German government early in World War II. He and his wife lost their young child to Lilith the child stealer in a tragic kidnapping. Years after his death it was proved by DNA testing that he had had a secret affair and a second family with a German woman for twenty years. Scorpio always has secrets.

Maurice Ravel (true Black Moon in Libra)

Maurice Ravel, a beloved French composer in the Impressionist movement, was well-known for his evocative *Bolero*, essential Lilith music. Give it a listen. Mean Black Moon in Scorpio was in the gifted grand cross of 1875. Saturn in Aquarius gave mastery of orchestration and form. Uranus in Leo shone on his Midheaven with creative originality and heart-grabbing musical richness. Pluto in Taurus tapped into the collective consciousness. His true Black Moon in Libra was opposite the impressionist planet Neptune in Aries. In his late 50s, possibly exacerbated by a car accident, he developed premature senility and memory loss, a mystery to physicans, perhaps a symptom of that subtle Black Moon-Neptune opposition. He is not known to have had any major love relationships. His mother was his closest friend until her death. "At heart, my only mistress is music," he said (www.maurice-ravel.net). He often found composing to be a grueling and exhausting endeavor.

Mistinguett

A kiss can be a comma, a question mark, or an exclamation point.
That's basic punctuation that every woman ought to know.

Born less than one month after Ravel, this queen of the Paris music hall was the most popular French entertainer of her era. With a limited but spectacular repertoire and equally spectacular, highly insured legs praised by Rodin, it was the force of her charismatic persona and risqué routines that made her reputation. Her Scorpionic explanation: "It is a kind of magnetism. I say,

'Come closer' and draw them to me.'" Said one-time lover Maurice Chevalier: "She had a way of moving which was the pinnacle of grace, but she was more than loveliness alone—she was Paris, the symbol of gaiety and good humor and courage and heart" (www.geocities.com/~jimlowe/mist/mistdex.html). Both her Black Moons were in Scorpio, just behind her Ascendant in that 1875 grand cross with Uranus in Leo (closest aspect), Saturn in Aquarius and Pluto in Taurus. The Liliths were also trine her Moon-Mercury in Pisces, enhancing her charisma and rapport with her audience.

Kiki de Montparnasse (true Black Moon in Libra)

In the next Paris music hall generation, Kiki, known as Queen of Montparnasse, performed in black hose and garters. She was a central figure in the vibrant Bohemian Montparnasse art scene of the 1920s. An artist herself, she posed for dozens of other artists, including photographer Man Ray (her companion for many years), Jean Cocteau and Alexander Calder, among numerous others. She owned a popular cabaret, Chez Kiki. Her memoirs, published in 1929, introduced by Ernest Hemingway and Tsuguharu Foujita, were banned in the U.S. until 1996. A daylily was named for her, as well as a high end Soho, New York boutique with a collection "that not only complements a life of romance and seduction, but helps to create it," proclaims the promotion at the Kiki online shop. Libra true Black Moon was quietly trine Pluto in Gemini, but mean Black Moon was conjunct the North Node on her Scorpio Ascendant, attended by Venus and Mars. Ooh la la!

Black Moon-Sun in Scorpio

Bill Wyman

This original Rolling Stones bass player (from 1962 to 1993) created the fretless electric bass and its unique sound. He was relatively level-headed for the rock world, eschewing drugs for sex, which led to some scandal, particularly his affair with a thirteen-year-old. Continually composing and doing other work through the years, Wyman began touring with his own Rhythm Kings in 1997. His books, *Rolling with the Stones* and *Stone Alone* chronicle his life with the band and without. At age seventy, he appeared in *Pirates of the Caribbean 2* as the father of Johnny Depp's quirky character. Wyman's Black Moon-Sun is opposite Uranus in Taurus and square his Moon in Aquarius.

Michael Landon

Born a week after Bill Wyman, this popular television actor was best known for his role of Little Joe on *Bonanza* and later for writing, directing and acting in *Little House on the Prairie* and *Highway to Heaven*. He contracted pancreatic cancer and died at age fifty-four.

Neil Young (true BM on Sagittarius cusp)

A classic rock and folk rock star from the 1960s, Young is still producing works in the early twenty-first century. Mainly a solo musician, he has played with major bands such as Buffalo Springfield; Crosby, Stills, Nash and Young; and Crazy Horse. He has recorded more than thirty albums, experiencing major ups and downs in his career as he experimented with different sounds. He was inducted into the Canadian Music Hall of Fame and the Rock and Roll Hall of Fame, and co-founded Farm Aid, which hosts music benefits for farmers, and the Bridge School Concerts, fundraisers for this innovative school for children with disabilities.

His sons both have cerebral palsy: son Zeke with Carrie Snodgrass (Oscar-nominated actress born two weeks earlier than Young also with Black Moon-Sun in Scorpio) and son Ben with wife Pegi. His daughter is epileptic, as is Young. His mean Black Moon Sun is in a T-square between Pluto in Leo on one side and the Aquarius Moon on the other. The Moon would be in the fifth house if he was born around 4:00 a.m. That time would also put a Jupiter-Chiron conjunction on his Ascendant. This seems likely.

His first son was born in 1972, the year of his Black Moon return that also saw his album "Harvest" hit the big time with the #1 single "Heart of Gold," and the same year one of his musicians overdosed on drugs. In 2005, Young had successful brain surgery for an aneurism. His album "Prairie Wind" came out afterwards, with lyrics that evoke the Scorpionic veils between worlds:

> *Late at night, lights dancing in the northern sky*
> *Like the Indian spirits trying to show me how to fly . . .*

Madeleine Korbel Albright

Albright was the first female U.S. secretary of state; she served under President Clinton, making her one of the highest ranking women in the history of the U.S. government. She also served as American ambassador to the U.N. Her Taurus Sun is opposite a Black Moon-Mars (retrograde) conjunction, reflecting her shrewd ability to strategize, to walk the corridors of male power, to face confrontation and at times controversy. Her father, Josef, a Czech diplomat, became the founding dean of the Graduate School of International Studies at the University of Denver. One of his favorite students was Condoleezza Rice. Following in her father's footsteps, Albright majored in political science and took U.S. citizenship. Married with three daughters, she co-founded the Center for National Policy and became a popular professor of international affairs and foreign policy at Georgetown University.

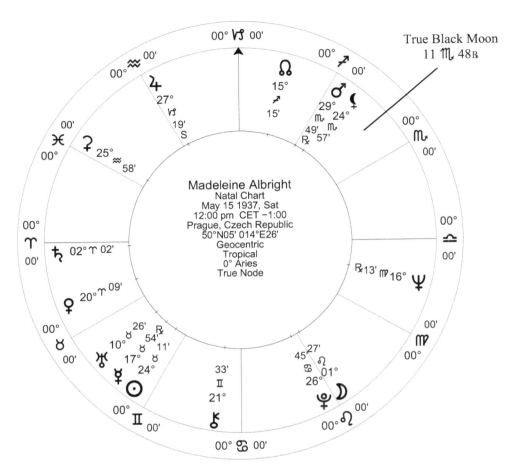

True Black Moon
11 ♏ 48ᴿ

Madeleine Albright
Natal Chart
May 15 1937, Sat
12:00 pm CET −1:00
Prague, Czech Republic
50°N05' 014°E26'
Geocentric
Tropical
0° Aries
True Node

Hugo Chavez

The popular president of oil-rich Venezuela is a dramatic personality, outspoken against U.S. policy. His Leo Sun is closely square mean Black Moon in Scorpio. He has spent hours on television, wooing his people with propaganda and music. He offers many social services, while amassing personal prestige. Once ousted from power, he came back and by example and endurance has helped lead South American resistance to U.S. pressure.

Black Moon-Moon in Scorpio

Roger Federer

In the top echelons of tennis and celebrity sports stars, Federer channels the intensity of his Moon-Black Moon-Uranus conjunction in the third house, all opposition Chiron in Taurus and

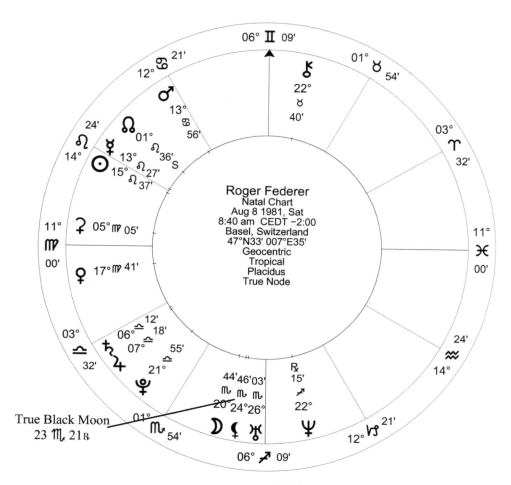

Roger Federer
Natal Chart
Aug 8 1981, Sat
8:40 am CEDT −2:00
Basel, Switzerland
47°N33' 007°E35'
Geocentric
Tropical
Placidus
True Node

True Black Moon
23 ♏ 21ʀ

square a Leo Sun. A class act of quiet intensity, high charisma and generous spirit, in his youth he had a fitful temper and has apparently done effective psychological work on himself—an ongoing Black Moon in Scorpio necessity. His family helps run his business, which includes a foundation that supports relief organizations in Africa and funds training for Swiss youth interested in sports.

During a Black Moon return in December 2007, he contracted mononucleosis, initiating a critical turning point in his career. His career took a downturn, publicly witnessed at the Australian Open in January 2008, during the exact week of the Black Moon return, when he failed to win this Grand Slam as expected. Other losses followed, including that of his extended top status. His Olympic gold medal win in men's doubles created a new feeling of winning for his country, not just himself, as he said. He went on to break more records in the tennis world, leading people to call him the best of all time. The Black Moon calls him to another and yet another Scorpionic test of will, desire and surrender.

Are you one or do you know one?

1954 Scorpios: I already mentioned this year as a profoundly Lilith one for those with the Sun in Aries or Cancer with Libra Black Moon in an unusually consciousness-shifting T-square dynamic. But Scorpios in the later months of this potent year might take the Lilith prize, as many of their Black Moon Suns are also conjunct Mercury, Venus (retrograde) and Saturn. Anyone born on October 26 or 27 wins the Lilith gold, with the New Moon in Scorpio as well!

More Black Moons in Scorpio

Mother Teresa. Kathleen Turner, Karl Pribram, Maya Angelou, Jacques Yves Cousteau, Emily Dickinson, Iris Murdoch, Jack Nicholson, Paris Hilton, Liza Minnelli, Kierkegaard, Kelsey Grammar, Paul Gauguin, Che Guevara, Dostoyevsky, Liberace

Black Moon in Sagittarius

Black Moon ranges broadly in this fire sign, with wide vision and mythic imagination, often on the international stage. Questions of philosophy, truth and morality loom large in the life, with unconscious personal beliefs imposing on agendas and being wrestled with in ways that may become evident for all to see. Sooner or later, the truth will out. These persons may live mythic lives, or become myth, even in their own lifetimes.

Black Moon-Ascendant in Sagittarius

Hans Christian Andersen (true Black Moon in Scorpio)

With mean Black Moon one degree from his Ascendant, this beloved writer is best known for his slightly dark fairy tales that remain classics beyond childhood. His poverty-stricken youth with mentally disturbed relatives left scars of inferiority. This famous Dane was an odd-looking person, socially awkward, phobic and sexually inhibited, yet also witty and sincere. Surely his exact conjunction of true Black Moon with Neptune in Scorpio contributes to both his psychological difficulties and also his fictional imagination.

Marion Woodman

"The Dark Goddess has to do with the Earth, the humus, the humility, the human . . . if she isn't brought to consciousness the planet is going to die."
—from interviews with Michael Bertrand at www.mwoodman.org/mw_interviews.html

One of the foremost teachers of women's psychosomatic spirituality and soul work, this woman

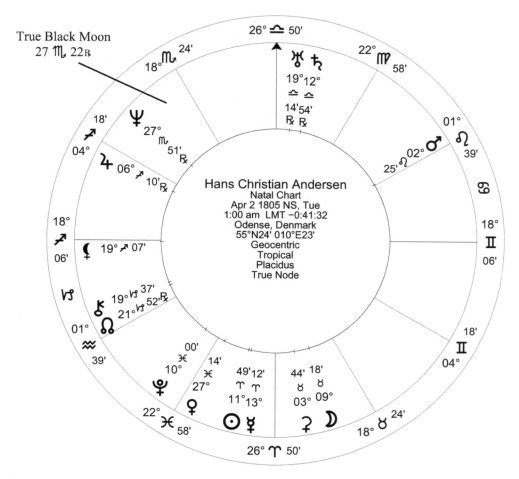

True Black Moon
27 ♏ 22ℝ

Hans Christian Andersen
Natal Chart
Apr 2 1805 NS, Tue
1:00 am LMT −0:41:32
Odense, Denmark
55°N24' 010°E23'
Geocentric
Tropical
Placidus
True Node

has clearly followed the path of the Dark Mother. Mean Black Moon at the last degree of Scorpio is square her close Mercury-Sun-Moon-Neptune conjunction in Leo. True Black Moon in mid-Sagittarius is conjunct an authoritative Saturn on the Sagittarius South Node-Ascendant, loosely but effectively aspecting her Venus in Virgo square Mars in Gemini.

Woodman started her career as a Jungian therapist. Her writings and experiential workshops tap into unconscious images that evoke healing, connecting ever more deeply with the dark goddess. She literally danced her way out of a wheelchair, swept onto the dance floor while listening to Polish party music. In her life and her work, she exquisitely expresses the empowerment and regeneration that come through the dark goddess and personal depth ecology. Her book titles reflect her journey. Early work with women with eating disorders addressed Moon issues, as published in *The Owl was a Baker's Daughter: Obesity, Anorexia Nervosa, and the Repressed Feminine, A Psychological Study*. Her writing became increasingly personal as she shed some of her academic persona and shared her own journey, as in *Leaving My Father's House: A Journey to Conscious Femininity. Dancing in the Flames* and *Coming Home to Myself: Reflections for Nurturing a Woman's Body and Soul*, continued to reflect her deeply transformational journey with both her Black Moon positions strong.

Jude Law

The likely early morning birth time gives this popular British actor a charismatic Sagittarius Ascendant colored by Black Moon, Neptune and Venus, which would go a long way in accounting for his charisma and high status on the "sexiest men alive" lists. Mean Black Moon conjunction Mars in Scorpio goes all the way. He has received Oscar award nominations for best actor and best supporting actor. His name resonates with a central meaning of this Black Moon sign, whichever "law" he follows. His birth chart from 1972 has Black Moon similarities with that of Brad Pitt, who was born a Lilith cycle earlier in 1963.

Black Moon-Sun in Sagittarius

Noam Chomsky (true Black Moon in Scorpio)

Either you repeat the same conventional doctrines everybody is saying,
or else you say something true, and it will sound like it's from Neptune.

This MIT professor of linguistics, philosophy and intellectual history generated ground-breaking theories and holds more than twenty honorary degrees from international institutions. He is equally known for his political activism. Mean Black Moon with Sun and Saturn is a strong, sturdy, intellectually-brilliant conjunction. He was the most often cited living scholar in the world from 1980 to 1992. Mercury closely conjunct the Black Moon heightens the intelligence and vision. A likely dawn birth puts this stellium on his Ascendant, and true Black Moon in Scorpio motivates deep investigations.

Rainer Maria Rilke (true Black Moon on Capricorn cusp)

I love the dark hours of my being.
  My mind deepens into them. . . .
Then the knowing comes: I can open
  to another life that's wide and timeless.—from *Book of Hours: Love Poems to God*

With Sun-Black Moon in fiery Sagittarius, this mystical poet of the modern age had a cosmic imagination that wrestled with angels and anxieties, inventing language and free-flowing images and metaphors. The Black Moons, both in Sagittarius, make a grand trine with Uranus in

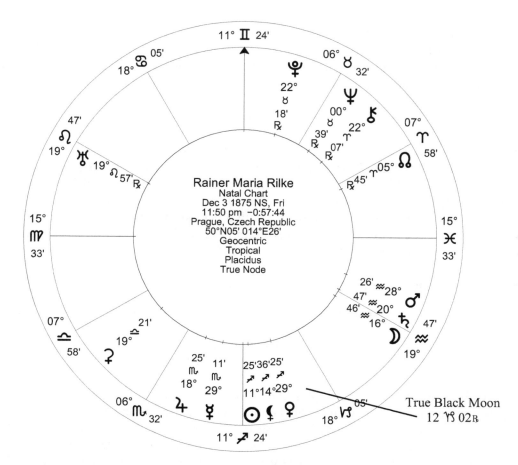

Rainer Maria Rilke
Natal Chart
Dec 3 1875 NS, Fri
11:50 pm −0:57:44
Prague, Czech Republic
50°N05' 014°E26'
Geocentric
Tropical
Placidus
True Node

True Black Moon
12 ♑ 02 ℞

Leo and Chiron in Aries. Mean Black Moon is close to his Sun, and true Black Moon is in the same degree as Venus, just over the Capricorn cusp. His mother dressed him as a girl for several years; then his father sent him to a military school. Such stark contrasts led him to develop illness as an escape to solitude. He traveled extensively, as Sagittarians tend to do, often hosted by wealthy patrons. He was lovers with an older Russian countess, Lou Andreas-Salome, who also had affairs with Nietzsche and Freud. He died at age fifty-one from leukemia as a result of an unhealed wound from a rose thorn. How mythic.

Frank Lloyd Wright

America's renowned master of organic architecture created designs to harmonize with the environment, leaving an indelible influence. True Black Moon was opposite the Gemini Sun with mean Black Moon closely opposition Mercury in its ruling sign. He developed the prairie house, a long, t-shaped structure emphasizing horizontal movement, and reconceptualized interior spaces, defining space by creating innovative, spacious and overlapping rooms. He also envisioned original designs for suburban community development. His most famous masterpiece, "Fallingwater," was built over a stream and waterfall. Black Moon Lilith exacted her due. His own home, "Taliesin" (named for the famous Welsh bard), was set on fire by a servant who murdered several people in the house, including Wright's beloved partner.

Are you one or do you know one?

Lilith has colored the lives of the following 1946 Sagittarians:

Sonia Gandhi (December 9): This Italian-born Indian politician became one of the most powerful women in the world. Her Sun is within a degree of true Black Moon and seven degrees from a precise mean Black Moon-Mars conjunction. Marrying into India's ruling Gandhi family, she was the daughter-in-law of Indira and the wife of Rajiv. These two Prime Ministers were both assassinated. After several politically unsettled years, Sonia Gandhi became a leader of the Congress Party, reviving its popularity, and then president of the Indian National Congress. She refused to become prime minister. The Black Moons in Sagittarius contributed to her extraordinary success as a woman and in a foreign country. She embraced and took to heart a very different philosophy and culture.

Jane Birkin (December 14): After starring in the 1967 movie *Blowup*, this singer, screen and stage actress. made her name in France in the provocative, sexually explicit French song and film, *Je t'aime . . . moi non plus*, for which she was nominated for the Best Actress Cesar Award. She has been awarded an OBE and Ordre National Du Merite in France. Active with Amnesty International, AIDS issues and immigrant welfare, she has traveled to such troubled areas as Bosnia and Rwanda to work mainly with children. Her Black Moon is conjunct Sun and Mars in Sagittarius as well as opposite Uranus in Gemini, any time of that day or night she was born. On her South Node, true Black Moon is square her Moon in Virgo.

Steven Spielberg (December 18): This most successful film maker directed some of the greatest movies ever made, using the mythic imagination of his close Black Moon conjunctions to Sun (mean Black Moon) and South Node (true Black Moon) to convey images and stories meaningful to the collective consciousness. Such movies as *E.T.*, *Close Encounters of the Third Kind*, *Raiders of the Lost Ark* and *Jurassic Park* are among these. Nominated for six directorial Academy Awards, he won three: in 1994 for directing and for best picture for *Schindler's List*, and in 1999 for directing *Saving Private Ryan*.

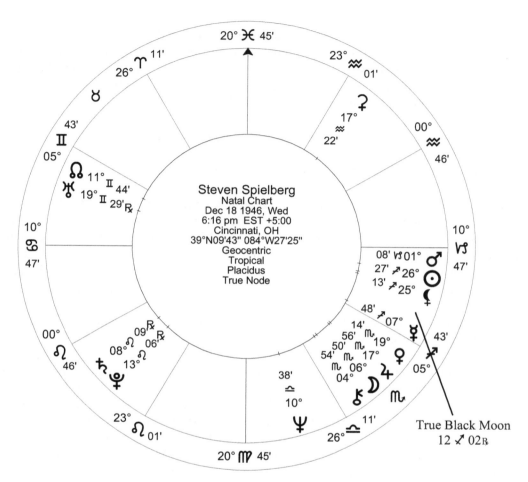

True Black Moon
12 ♐ 02℞

Black Moon-Moon in Sagittarius

Ted Bundy (November 24)

This notorious American serial killer brutally raped and murdered dozens of women. Born about a week earlier than Gandhi, Birkin or Spielberg, it was his Moon in later Sagittarius rather than his earlier Sun that was conjunct Black Moon, Mars and the South Node and opposition Uranus in Gemini. Born in a home for unwed mothers, he was left for two months until taken in by his violent grandfather, whom he idolized. Highly intelligent and proud of it, this charismatic all-American looking man led a double life, dating and working in politics. He kept his high volume homicidal life well hidden until his mental instabilities got beyond his control. After being on death row for ten years, he was executed in 1989.

Anthony Hopkins

Born in a different year, but also with Moon in Sagittarius conjunct the two Black Moon positions, this famed actor starred in the role of a serial killer, the very dark, infamous Hannibal Lector. One of Hopkins' favorite movie roles was as the real life character Burt Monroe in *The World's Fastest Indian*, a very Sagittarian story. (See my book *Living Lilith: Four Dimensions of the Cosmic Feminine* for a fuller profile of Hopkins.)

More Black Moons in Sagittarius

Deepak Chopra, Claude Monet, Gaston Bachelard, George W. Bush, Yoyo Ma, Dan Brown, Modigliani, Tracy Chapman, Odillon Redon, Gianni Versace, Bruce Willis, Keanu Reeves, Robert Redford, H.G. Wells, Jean Genet, Dante, Bill Gates, Russell Crowe, Bill Clinton, David Bowie, Juliet Binoche, Cyrano de Bergerac, Cher, Verdi, Jack London, Pope Jean-Paul II, Isabella Adjani, Aleister Crowley, Nostradamus, William Joseph Turner, Martin Luther King

Black Moon in Capricorn

Social forms, structures and systems are deconstructed, reformed and reframed by Lilith in Capricorn. These people tap into the basic principles and subatomic grids of Nature and human nature, creating new genres. Father and authority figure issues can play into their lives in particularly profound ways. Control, capability and contemplation become means to self-measurement and self-mastery.

Black Moon-Ascendant in Capricorn

Marie Curie

This famous scientist worked with her husband investigating chemical structures, isolating radium and ultimately dying from its poisonous effects. Winner of two Nobel Prizes, one in physics and one in chemistry. she was both the first woman to win the award and the first person to win two. Like Kovalevskaya (below), Curie faced obstacles as a woman in the world of science, denied entrance into the French Academy because of her gender. Scientist and mother, she and her husband lived a rich family life. After his sudden, accidental death, she caused a scandal by having an affair. The Black Moons sandwiched her Ascendant and actively engaged with two outer planets, square Neptune in Aries and opposition Uranus in Cancer on the seventh house cusp. In Capricorn style, the scientific tradition was passed on to their elder daughter, who also won a Nobel Prize for chemistry.

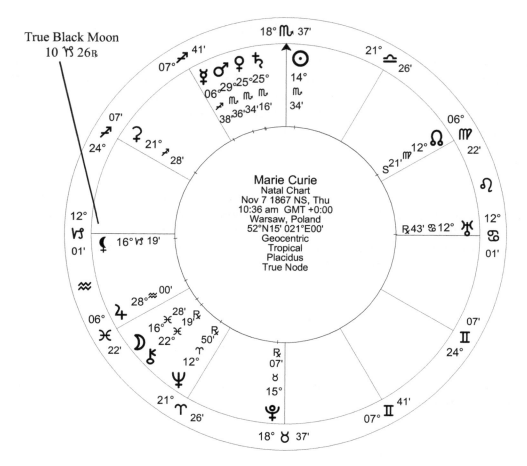

True Black Moon
10 ♑ 26ʀ

Marie Curie
Natal Chart
Nov 7 1867 NS, Thu
10:36 am GMT +0:00
Warsaw, Poland
52°N15' 021°E00'
Geocentric
Tropical
Placidus
True Node

Hiraoki "Rocky" Aoki

This former Olympic wrestler and off-shore powerboat racer opened the Benihana chain of Japanese steakhouses and hired entertaining chefs who juggled knives. He once faced deportation from the U.S. for tax evasion and was also investigated for insider trading, at which time he pulled back from the business. He is the father of supermodel Devon Aoki and Steve Aoki of Don Mak Records. His Black Moon-Ascendant was in an entrepreneurial cardinal cross with Sun-Mercury in Libra, Moon-Saturn in Aries and Chiron in Cancer, and also made a grand trine in earth signs with Uranus in Taurus and Mars-Neptune in Virgo.

Mel Gibson

With mean Black Moon Lilith widely conjunct his Sun and closely square his Moon, this popular Irish-American-Australian actor has a strong Lilith signature. His name comes from two

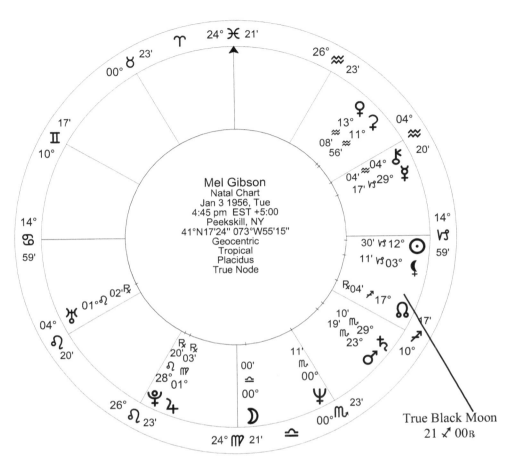

Mel Gibson
Natal Chart
Jan 3 1956, Tue
4:45 pm EST +5:00
Peekskill, NY
41°N17'24" 073°W55'15"
Geocentric
Tropical
Placidus
True Node

True Black Moon
21 ♐ 00ʀ

Irish saints: Saint Mel, for whom he is named for the Irish Saint Mel, the founder of his Irish mother's diocese. Although he was born in the United States, his father moved the family from New York to Australia, the ancestral land on that side of the family, when Mel was age twelve.

Gibson has been highly successful on stage and in film in a range that includes action-adventure films, historic drama, and romantic comedy, from the *Mad Max* and *Lethal Weapon* series to *The Bounty*, *Gallipoli* and *Braveheart* (winner of the Best Picture Oscar) to *Hamlet*, he has captured the screen with his charismatic presence. He continues to work on stage and has branched out to award-winning directing and producing. His Capricorn business interests include a portfolio of property investments and substantial contributions to various charities and good works organizations.

Gibson has the quirky, often raunchy, Capricorn humor for which he is known in film and on television. The Black Moon Sun square Moon shows up; in issues with his father, as well as in his own fathering of sevel children during a nineteen-year span in his marriage and a more recent child with a new woman. He has struggled mightily with alcoholism and has become personally involved in helping others.

The true Black Moon on his North Node in Sagittarius also has an impact as he grapples with profound moral questions of, and an intense involvement with, the early Catholicism that led to his production of *The Passion of the Christ*. This position gives him a mythic, storytelling imagination similar to that of Steven Spielberg. A typical Sagittarius outspokenness has invited labels of anti-feminism and anti-Semitic. He is fearless about giving his opinion on any issue.

Black Moon-Sun in Capricorn

Jean of Luxembourg

His Royal Highness Grand Duke Jean Benoît Guillaume Robert Antoine Louis Marie Adolphe Marc d'Aviano of Luxembourg reigned from 1964 to 2000. With a long list of titles, awards and decorations, his close Black Moon-Sun in Capricorn carries on a long royal tradition. True Black Moon is closely conjunct Mercury, square Chiron in Aries and opposite Pluto in Cancer, making a powerful T-square. When Germany invaded Luxembourg in WWII, the Grand Ducal family escaped first to France. Jean went on to Canada. He volunteered in the British Army, participated in the Normandy landings and helped liberate his country. In 1953, he married Belgian princess Josephine-Charlotte, a Lilith lady herself with Black Moon on her Sun (see Black Moon in Libra section). In addition to their mutual Black Moon-Sun squares, the chart of the royal marriage shows Chiron in Capricorn on the Grand Duke's Sun-Black Moon. The power of European ropyalty united them and is carried on by their five children. Jean abdicated the throne to his son Henri in 2000.

Gustave Dore (mean Black Moon in Sagittarius)

A master engraver, sculptor and painter, Dore famously illustrated classic works, such as *The Divine Comedy, The Rime of the Ancient Mariner, Perrault's Fairy Tales* and an English Bible. A project of several years and ten engravings, *London, A Pilgrimage*, was a success but garnered criticism from some quarters because of its honest portrayal of London's common and "vulgar" poverty-stricken areas. A child prodigy, he remained close to his mother, never married, and died at age fifty-one of heart failure. It was true Black Moon conjunction his Sun and in a grand trine with Saturn in Virgo and Chiron in Taurus that contributed to his mastery in the art of illustration. His mean Black Moon in Sagittarius loved the storytelling in art.

Eddie Vedder (true Black Moon in Sagittarius)

Lead singer and guitarist of the grunge band Pearl Jam, Vedder took his mother's maiden name in his late teens when he found out the man he thought was his father was actually his stepfather (Capricorn father issues). He plays a number of instruments and has performed with a long list of famous musicians. Politically outspoken in a way that sometimes seems to undermine the band's success, he discusses politics during shows and sports an Earth First! tattoo. Pearl Jam was a headliner for the U.S. Vote For Change tour in 2004. Vedder is a surfer and vegetarian. His mean Black Moon is conjunct his Sun and trine his Moon. True Black Moon, back in Sagittarius, is also active, conjunct Mercury and square Mars. This aspect indicates his outspokenness. Add the Pluto-Uranus conjunction of the 1960s decade to that Mars square Black Moon and you get grunge!

Sofia Kovalevskaya

A mathematical genius at a young age, this woman became the first to hold a full professorship in a European university (in Sweden). Her father (Capricorn father influence) had wallpapered her nursery with pages of calculus notes, which caught her early interest. Listening to her uncle talk on mathematical theories struck her imagination and instilled a reverence for mathematics as a mysterious doorway to a world beyond the ordinary. Unable as a woman to pursue her education in Russia or to travel alone, she made a marriage of convenience to be able to study abroad. Yet even in Germany she had to be privately tutored by famous mathematics professor Karl Weierstrasse because women weren't allowed at the University of Berlin. Not surprisingly, she became an advocate of women's rights.

Kovalevskaya was eventually awarded a Ph.D. from the University of Gottingen. Her winning entry for the Prix Bordin from the French Academy of Science in 1886 was so good that the committee increased the prize. She wrote "On the Rotation of a Solid Body About a Fixed Point," developing an original theory about the movement of nonsymmetrical bodies. She wrote a number of works offering groundbreaking theories in math, mathematical physics and authored literary pieces as well. Achievement and death came together in her life. She died of pneumonia at age forty-one.

This extraordinary woman has outer planet aspects to Capricorn Black Moon as does Marie Curie, but this time it is Uranus and Pluto in Aries and mean Black Moon conjunction her Sun. True Black Moon is more closely conjunct Venus, reflecting her struggle and eventual success as a woman in science.

Leonardo Sciascia

This major modern literary figure wrote about the socio-political flavor of Sicily and the shadowy power of the Mafia. His interest in unsolved crimes and political mysteries, historic and contemporary, led to an original work of fiction and landmark works of Italian literature. A communist, he was also active in the Radical Party in both the Italian and European Parliaments. His chart shows a Capricorn sequence of Moon, Mercury, Sun and Black Moons, all at the top of his chart and square his Aries Ascendant. Sciascia was born on the day of a very potent Lilith corridor, with true Black Moon in the same degree as the Sun and mean Black Moon two degrees later.

Samantha Mumba (true Black Moon in Aquarius)

Born of an Irish mother and Zambian father, this Grammy-nominated singer was a 2000 chart topper in Ireland and the UK with her song, "Gotta Tell You." She made a memorable Lilith appearance wearing a $9 million dollar Spider Web Diamond Dress at the premier of Spiderman II, as befits a very close Sun, Black Moon, Mercury conjunction.

Black Moon-Moon in Capricorn

Anais Nin

This French fiction writer is especially known for her provocative *Diaries*, which give some real-life background to her erotic, psychologically penetrating and surreal scenarios of relationship, love and desire. Conjunct Moon at the nadir, Black Moon suggests deeply disturbing family and home issues. Her father left the family early on, but returned into her life in an incestuous relationship. A character named Lilith appears in several of her stories. In "The Voice," Lilith is a kind of alter ego, a little girl self engaged in psychological analysis with a father figure and in dialogue with her woman self.

Are you one or do you know one?

From May 1991 through late January 1992, the Black Moon joined the rare Uranus-Neptune conjunction in Capricorn. The effect was felt strongly by Cancers in summer 1991 and by Libras that fall and was very potent for Capricorns the following winter, half of whom also had Mercury and Mars there. This sacred geometry conjunction offered unusual transfigurative energy.

More Black Moons in Capricorn

Leonardo DaVinci, Jean Renoir, Auguste Renoir, Arnold Schwarzenegger, Hilary Clinton, Camilla Parker Bowles, Tom Hanks, Elton John, Grace Kelly, Kofi Annan, William Lilly,

Timothy Leary, Stephen King, King Hassan II of Morocco, Bill Maher, George Patton, Eva Braun, Ray Bradbury, Shri Meher Baba, Joseph Smith, Pablo Cassals, Mata Hari

Black Moon in Aquarius

Outstanding individuality is common among this uncommon assortment of people. They may make an indelible dent in the social fabric, yet Lilith requires stepping back to maintain an objective perspective about themselves and their roles. Black Moon in Aquarius is plugged into a large energy grid, downloading input and ideas beyond common understanding. They may feel like strangers in a strange land, or the odd one out, seeking connection through community even while feeling personally isolated.

Black Moon-Ascendant in Aquarius

Jimmy Hoffa (true Black Moon in Pisces)

Rising to the powerful position of president of the Teamsters' Union, in which he served from 1957 to 1971, this labor leader had purported ties to the Mafia and organized crime. He was convicted for attempted bribery, but President Nixon commuted his sentence. He mysteriously disappeared in 1975, and is presumed murdered. He has become a popular icon, referred to in various artistic contexts. Black Moon was conjunct his Ascendant, Sun and Mercury, all between 25 and 28 Aquarius, in square to an exact Moon-Saturn conjunction in Taurus. True Black Moon is conjunct Chiron in the first house, adding to his aura and the mystery.

Robert Louis Stevenson

Author of such classics as *Robinson Crusoe, The Strange Case of Dr. Jekyll and Mr. Hyde, Treasure Island* and *A Child's Garden of Verses*, Stevenson's Black Moon Ascendant was square Sun, Mercury and Mars in Scorpio. Although of delicate health, he traveled extensively. He married an American woman, lived in the U.S. and sailed the South Pacific for several years, during which time he became friends with the Hawaiian royal family. Eventually he bought land and settled in Samoa with his wife, working his estate and becoming involved in local affairs. He died there at age forty-four, after living a short but colorful life. "Sick and well, I have had a splendid life of it, grudge nothing, regret very little . . . would hardly change with any man of my time."

Alfred Dreyfus

With Black Moon conjunct the North Node and Chiron on his Ascendant, this man gave his name to the Dreyfus Affair, a major political scandal at the turn of the twentieth century. One of a few French military men of Jewish ancestry, he was falsely accused of treason and imprisoned

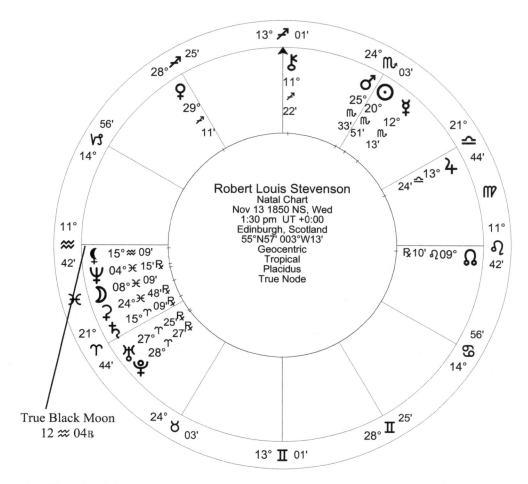

True Black Moon
12 ♒ 04ᴿ

on Devil's Island for some years before he was pardoned and later exonerated. He went on to serve in World War I and was made a knight in the Legion of Honour. This affair has been the subject of continuing interest and numerous dramatic depictions. The mean Black Moon-Chiron conjunction is at 18Aquarius. The Sabian symbol reads: "At masquerade the last man unmasks, urged by the girls." Dane Rudhyar's short form interpretation indicates a relevant theme in this case: "The introvert's desire to protect himself from social judgment."

Black Moon-Sun in Aquarius

Alice Cooper (true Black Moon in Capricorn)

This "shock rock" star is one of the most enduring and influential musicians of his generation. His Lilithian "Only Women Bleed," "No More Mr. Nice Guy" and other songs are classic staples of rock radio. His Sun-Black Moon opposition to Saturn-Pluto in Leo is an indication of his high theatrics and artistic longevity (helped along by Neptune rising). Hard to classify, this individual has also been a film actor, golf celebrity and restaurateur, suggesting the pull of the Capricorn true Black Moon. This may also account his long-term marriage and growing up as a preacher's son.

Isabelle Eberhardt

An extraordinarily liberated yet contradictory woman for the late nineteenth century, she was an adventurer, journalist and self-described vagrant who journeyed extensively in North Africa dressed as a man. An illegitimate child of an aristocratic Swiss woman, she rejected conventional Western values for a free lifestyle; Eberhardt became a Moslem and was initiated into a Sufi sect, yet relished drugs, drink and sex. She married an Algerian soldier and died mysteriously at age twenty-seven, a victim of murder or a flash flood. Her evocative short stories, such as "The Oblivion Seekers," "Outside" and The Breath of Night," describe hashish smokers, holy men and hopelessness, the back alleys of Algiers, and the power of desert sounds. Her individualistic Aquarian Sun, already opposition Uranus in Leo and square Pluto in Taurus on her Ascendant, was pushed further into the wilderness by the Sun's conjunction with mean Black Moon. True Black Moon was closely conjunct Venus and Mercury, all square Neptune in Taurus.

Johann Petursson (true BM in Pisces)

A delicious and unusual example worthy of Aquarius, Johann Petursson, called The Icelandic Giant, was seven feet, seven inches high, one of the tallest men in the world. He appeared in *Prehistoric Women,* among other movies, and in magazines, circuses and carnivals. He joined Ringling Brothers in 1948. True Black Moon in Pisces conjunction Chiron may indicate some personal sensitivities and problems, but helped make his oddity worthy of the film and circus industries.

Sonny Chiba

With Black Moon so close to his Sun and opposite Pluto in Leo, this placement no doubt contributed to his unusual path in life as one of the first stars, action directors and teacher-trainers of martial arts movies. In his forty-year career he was influential in defining the genre. *The Street Fighter* was one of his most famous roles.

J.K. Rowling (true Black Moon in Capricorn)

The author of the Harry Potter series has Black Moon opposite her Leo Sun. Her creative work connects with the collective consciousness of the times, with multigenerational fans around the world. The true Black Moon in Capricorn gives her stature and places her in a continuing literary tradition, while taking it to a new level. Wildly successful, she is big business. The Capricorn magic of Harry Potter took her from being a single mom on welfare status to becoming one of the wealthiest woman in Britain. Her home is a castle.

Black Moon-Moon in Aquarius

Anton Pieck

This popular Dutch painter, graphic artist and master art teacher has Black Moons closely conjunct the Moon and square Uranus. He developed a unique style, often fairy-tale like, with childlike whimsy yet strong detail. He is most famous for designing Efteling Fairy Park, a national park that is Holland's equivalent of Disney Land. In a 2004 survey, Pieck was listed in the top 100 greatest Dutchmen.

Edgar Cayce

America's best-known psychic had a Black Moon corridor spanning 7-22 Aquarius. The mean Black Moon is square his Moon and Neptune, appropriate for this man, called the Sleeping Prophet. The true Black Moon is closely square Pluto. While in a trance this minister could give psychic information about the lives and past-lives, health and issues of persons even at a distance. His vast array of esoteric information continues to be catalogued and studied by The Edgar Cayce Foundation, a world-renowned study center in Virginia Beach, Virginia.

Notable Black Moon in Aquarius

Jacques Derrida

If this work seems so threatening, this is because it isn't simply eccentric or strange, incomprehensible or exotic . . . but competent, rigorously argued, and carrying conviction in its reexamination of the fundamental norms and premises.
—from *Points. . . . Interviews 1974-1994*

The Black Moon in Aquarius is not a dynamic stand-out in his chart, with only a quincunx to a *cazimi* Sun-Mercury in Cancer and a square to Chiron. Perhaps that *cazimi* quincunx with true Black Moon heightens the work of this influential modern philosopher that so well expresses the high intelligence of this Black Moon placement.

I was compelled to dig into his chart. It shows a close-to-perfect grand trine formed by true Black Moon at 29 Aquarius, Pallas Athena in Gemini and the South Node in Libra, giving him a rare theoretical perspective. Known as the founder of deconstruction, Derrida's controversial ideas and extraordinary analyses of the nature of language, writing, and meaning have had a profound and continuing effect on modern philosophy and literary theory. He was committed to

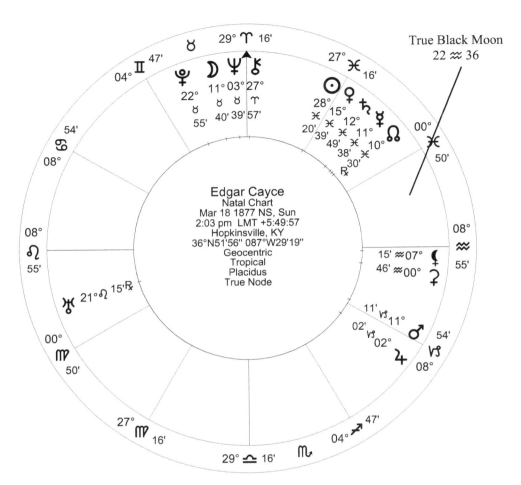

the cause of social justice (very Aquarian) and was politically active. Certainly the rest of his chart, in particular a close Sun-Mercury-Pluto conjunction in Cancer square Uranus in Aries, supports the nature of his work.

Are you one or do you know one?

Rare souls arrived in January 2001 to embody a Neptune-Black Moon conjunction in Aquarius. There must have been some strong solar flares that month! Some late Capricorns had Mercury, and the Aquarians had the Sun plugged into the rarefied conjunction. January 24 New Moon babies had Sun-Moon at five degrees and the Neptune-Black Moon at 7 Aquarius. Let's keep our eyes on these angels in human form and help them remember who they are.

More Black Moons in Aquarius

Elwood Babbitt, Martine Navratilova, Leo DeCaprio, Sean Connery, Bjork, Clint Eastwood, Cary Grant, Louis XIV, Jon Voight, Neil Armstrong, Georges Seurat, Pope John Paul I, Alexandra David-Neel, Loreena McKennet, Chris Rock, B.F. Skinner, Joseph Campbell, Paul Ricoeur, Savonarola, Anton Chekhov, Germaine Greer, George Balanchine, Fritjof Capra, E.E. Cummings, Mikhail Baryshnikov, Alanis Morissette

Black Moon in Pisces

Elusive shape shifters, these people often have a loose hold on reality and high, sometimes unattainable, ideals. Pisces can either catch the wave and ride it high, or be pulled under—sometimes both in one lifetime. The martyr tendency of Pisces is usually lived out in some way as an aspect of sacrifice or surrender. Extraordinarily sensitive, their poetic souls tune into rare wavelengths and thrive on beauty.

Black Moon-Ascendant in Pisces

Phyllis McGuire (true Black Moon in Aquarius)

Phyllis was the youngest of the three McGuire Sisters, the most popular American female vocal group of the 1950s, best known for their song "Sugartime." These Midwestern females hit the big time and had connections with Sinatra and JFK. Phyllis had an affair with Mafia boss Sam Giancana, who liked to hang out with famous entertainers (he had Black Moon in Leo). Enjoying a rich life, she later became a Las Vegas philanthropist. Mean Black Moon was on her Pisces Ascendant (and square Neptune in Virgo), and true Black Moon was in the twelfh house conjunction her otherwise unaspected Aquarian Sun.

Black Moon-Sun in Pisces

Drew Barrymore

Born into a multi-generational acting family, Drew entered the acting profession early. Her first appearance was in a commercial when she was eleven months. She hit stardom at age seven in the movie *E.T., The Extra-Terrestrial.* A troubled youth of drugs and drink is described in her autobiography, *Little Girl Lost,* written when she was twenty-five. Lost girl turned into bad girl sex symbol when she had nude scenes in a series of films. In the 1990s she changed tone again, aspiring to produce movies as well as star in them—all of this before her Saturn return. The shape-shifting Sun-Black Moon in Pisces on her Midheaven plays strongly in her career.

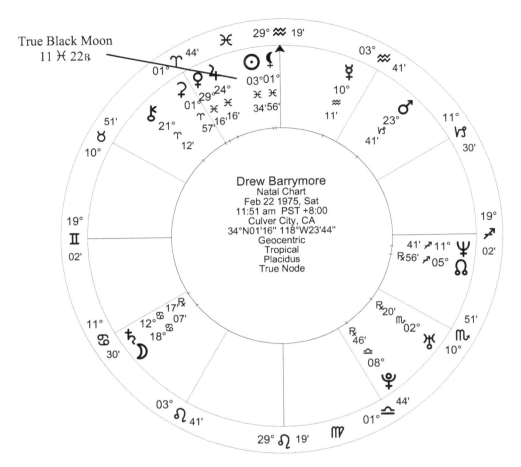

Mikhail Gorbachev

An inscrutable and charismatic politician with a signature birthmark on his forehead, Gorbachev has a close Sun-Black Moon conjunction. Leader of the Soviet Union from 1985 to 1991, he transitioned his country out from behind the Iron Curtain toward democracy and the modern world, winning the Nobel Peace Prize in the process. Communism crumbled and the Soviet Union dissolved. His tenure ended in a coup. *Time* magazine named him as one of the top one hundred most important people of the twentieth century.

Yitzhak Rabin

I have always believed that the majority of the people want peace and are ready to take risks for peace.—from last speech at peace rally where he was assassinated, November 4, 1995

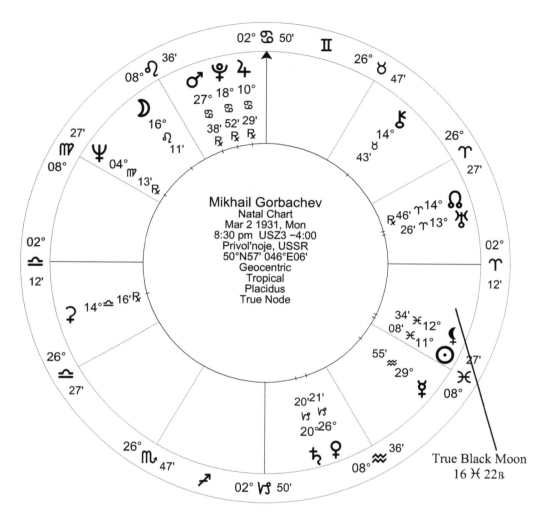

Mikhail Gorbachev
Natal Chart
Mar 2 1931, Mon
8:30 pm USZ3 −4:00
Privol'noje, USSR
50°N57' 046°E06'
Geocentric
Tropical
Placidus
True Node

True Black Moon
16 ♓ 22ᴿ

The first native-born prime minister of Israel, Rabin was in office twice: in the 1970s and again in the 1990s. Starting in the military, he became chief of staff in the Israel Defense Forces and was a member of the elite strike force of the underground Haganah. He served as ambassador to the U.S. As Prime Minister he helped engineer the Oslo Accords, for which he received the Nobel Peace Prize in 1994, along with Yasser Arafat and Shimon Peres. This act was both lauded and condemned in his country. A radical dissident assassinated Rabin in 1995, shocking the nation and the world. His last speech was given at a peace rally in front of about one hundred thousand people. His Black Moon corridor from 6 to 16 Pisces encompassed Uranus, Sun and Venus.

Jack Kerouac

All human beings are also dream beings. Dreaming ties all mankind together.
—from *Book of Dreams*

Born the same month as Rabin, Kerouac's Black Moons also engaged Uranus, Sun and Venus, with the additional impact of a Virgo Moon opposition. This man of French-Canadian parentage became one of the most important American writers, crisscrossing and famously portraying the changing landscape of America. His 1957 novel, *On the Road*, was a "text" for the Beat generation. Restless for meaning, he heralded the 1960s, experimenting with drugs and traveling the world. Alcoholism led to an early death at age forty-seven. He said, "I hope it is true that a man can die and yet not only live in others but give them life, and not only life, but that great consciousness of life."

Black Moon-Moon in Pisces

Susanna Kaysen

Author of *Girl, Interrupted,* made into a 1999 movie, Kaysen writes and lectures about her late teen years in Harvard's private psychiatric institute, diagnosed with a borderline personality disorder. Continuing her Black Moon-Moon in Pisces saga, her 2001 book, *The Camera My Mother Gave Me,* chronicles the medical journeys of her dysfunctional vagina, while deeply exploring issues of female sexuality, desire and desirability.

Elizabeth Scarlett Jagger

Mick Jagger's daughter, with model Jerry Hall, has both Black Moons closely aligned with her Mercury-Moon-Sun conjunction in the fourth house, a clear indication of Black Moons in the family. Father is listed in the Leo section with Sun-Black Moon; mother has Black Moon in Capricorn square the Moon. Elizabeth is a fashion model and actress; she started out by walking the runway with her mom.

Robert Plant

Although the time of birth is not confirmed, the Moon is almost certainly in Pisces conjunct Black Moon Lilith in the chart of this famed rock musician. It certainly "feels" right that this lead singer of the 1960-1970s heavy metal, hard rock band Led Zeppelin would have such a conjunction. As a further confirmation of the compassionate Pisces placement, and his elevated Mercury in Virgo opposition that seeks to be of practical service, Plant has given multiple benefit concerts for Armenian earthquake relief and children's cancer clinics.

Notable Black Moon in Pisces

Gloria Estefan (true Black Moon in Aquarius): This queen of Latin pop has won five Grammy Awards in her superstardom as a Cuban-American singer-songwriter. With the Miami Sound Machine (true Black Moon in Aquarius collaboration), she created a rhythmic fusion of pop, disco and salsa that crossed cultures and borders. Her chart has Sun-Mars in Virgo opposition Black Moon in Pisces T-squared by a Moon-Saturn conjunction in Sagittarius. The Sun-Black Moon opposition is the strongest part of that dynamic. After recovering from a near fatal accident on March 20, 1990, followed by spinal fusion, she made a world tour comeback with her "Into the Light" album that featured the hit song "Coming Out of the Dark." No doubt, she did. On the day of the accident, Pluto was conjunct the Black Moon close to her Scorpio North Node.

Are you one or do you know one?

1966 Scorpios and Sagittarians, born at the height of the 1960s, notch it up another level with a Black Moon-Saturn conjunction in Pisces opposition the rare Uranus-Pluto conjunction that made the 1960s "The Sixties." Some charts add Mars in Virgo or Venus in Sagittarius to the mix for some intense emotional dynamics tending toward the tantric.

More Black Moons in Pisces

Tina Turner, William Shatner, Angelina Jolie, Gerard Depardieu, Halle Berry, James Dean, Andre Breton, Rubens, Paul Eluard, Cyd Charisse, Al Jarreau, Talleyrand, Herbie Hancock, Sir James Barrie, Leonard Nimoy, David Beckham, Kiefer Sutherland, Yusuf Islam (Cat Stevens), Albert Camus, Christian Dior, King Louis XV, Janet Jackson, Judy Garland, Queen Margrethe II of Denmark

Black Moon Beatles

The Beatles became an ultra-iconic quartet, a powerhouse of the "English invasion" of the 1960s revolutionary world of drugs, sex and rock-and-roll. Together, then individually, four musicians explored and forged new dimensions of popular music both technically and conceptually, to create an unforgettable legacy that has become embedded in our collective consciousness.

The song-writing partnership of Paul McCartney and John Lennon produced the major percentage of The Beatles' songbook. These two came up early in my Black Moon research. Lennon has the Black Moon conjunct his Ascendant, McCartney conjunct his Sun. As I looked further into the group Liliths, I found that, indeed, Ringo Starr and George Harrison also have significant Black Moons placements, strong enough to make the group worthy to be called the Black Moon Beatles. Let's see why.

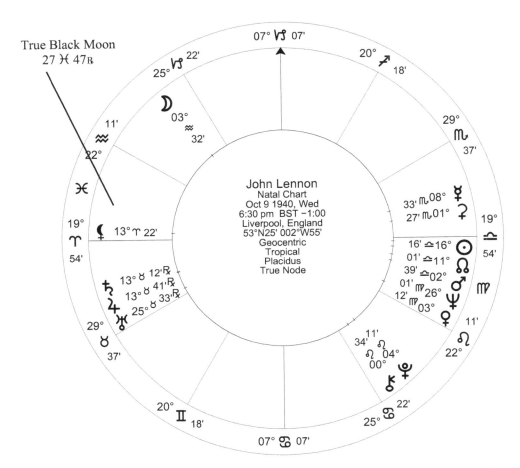

True Black Moon
27 ♓ 47ℝ

John Lennon
Natal Chart
Oct 9 1940, Wed
6:30 pm BST −1:00
Liverpool, England
53°N25' 002°W55'
Geocentric
Tropical
Placidus
True Node

Interestingly, all four of the band members have mean and true Black Moons in different signs. That adds complexity and nuance to the expression of this subtle energy and offers a good opportunity to further consider such dual placements.

John Lennon (mean Black Moon, 14 Aries;
true Black Moon, 28 Pisces; 16-degree corridor)

A leading voice, both as vocalist and public front man for the group, Lennon pushed creative frontiers throughout his life. He wrote and produced dozens of well-known songs. In addition, he published art work and witty, poetic writings. He was politically outspoken and active.

Mean Black Moon is conjunct Lennon's Aries Ascendant, giving him that strong, sometimes feisty, charismatic personality. It is also conjunct the South Node and opposite his Libra Sun-North Node, all powerful aspects. The Black Moon opposite his Libra Sun reflected powerful persona and the strong, sometimes confrontational, partnerships in his life. His first wife Cynthia is the mother of his son Julian Lennon. His second marriage to Japanese artist Yoko Ono expedited the break-up of The Beatles and led to his solo career. They had one miscarriage before son Sean Lennon was born. John had major enemies. Due to his outspoken political opinions, he became the target of an FBI investigation that sought to deport him, considering him a threat to Nixon's continuing administration,. He was shot by a former fan in 1980.

Lennon's true Black Moon was in watery Pisces, intercepted in the twelfth house. He had a quiet, visionary and spiritual side, which he revealed to some extent in the post-Beatles time in his life when he stepped away from the music machine. He and Ono famously staged a publicized sleep-in, a radical way to say "Give Peace a Chance." The inscrutable nature of the twelfth house enhances the subtle nuances of Black Moon, making it difficult to discern the scope of this placement. There is often a sense of sacrifice in Pisces, certainly of his privacy, and also an aspect of the martyr expressed in his death. One of his most enduring solos, "Imagine," has a Pisces theme:

> *"You may call me a dreamer, but I'm not the only one.*
> *I hope some day you'll join us, and the world will be as one.*

Paul McCartney (mean Black Moon, 23 Gemini;
true Black Moon, 6 Cancer; 13 degree corridor)
Sir Paul wrote scores of songs, both with and without Beatles partner John Lennon. Post-Beatles, he formed the group Wings with wife Linda, who died of cancer at age fifty-six. A typical Gemini of many talents, he has also composed classical works. His mean Black Moon is on the Midheaven, sandwiched between his Gemini Sun and ruling planet Mercury for a close triple conjunction. Mean Lilith is also square Neptune on his late Virgo Ascendant. This suggests rare skills in music and points to the excesses indulged in by all the Beatles in the drug-filled rock world. His self-perception may be confused or clouded by his public image. Paul's multiple square is in close dynamic aspect to John's true Black Moon in Pisces, reflecting their close companionship and friction.

Geminis tend to maintain a youthful look, as Paul has, and can sometimes be accused of being superficial, writing "another silly love song. . . . What's wrong with that?" Another obvious expression of Gemini's theme of polarity was his famous duet with Michael Jackson, "Ebony and Ivory."

> *Ebony and ivory live together in perfect harmony*
> *Side by side on my piano keyboard, oh lord why don't we?*

His many songs have been sung and resung countless times by so many many others. Paul has been an outspoken advocate for vegetarianism, animal rights, breast cancer awareness, mari-

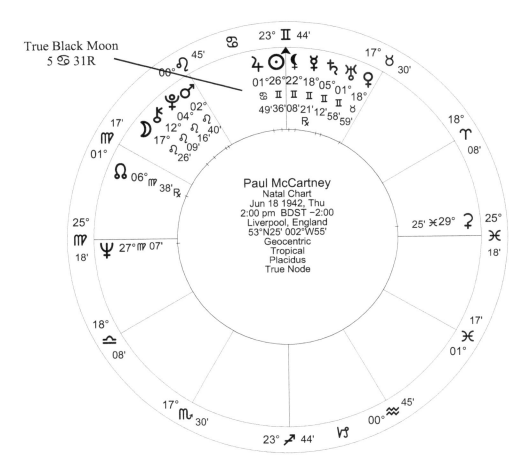

True Black Moon
5 ♋ 31R

Paul McCartney
Natal Chart
Jun 18 1942, Thu
2:00 pm BDST −2:00
Liverpool, England
53°N25' 002°W55'
Geocentric
Tropical
Placidus
True Node

juana legalization and other issues.

Paul's true Black Moon is in Cancer conjunct an exalted Jupiter in the tenth house, still close to his late Gemini Sun. This dual Black Moon adds complexity to its expression. With the Gemini leaning into Cancer, Paul could sing those soulful, soft Cancer-like ballads, as well as rollicking rock. Cancer in the tenth house has a finger on the pulse of what the public needs.

It is also the sign of a family man. A first engagement in the early days of the Beatles ended after his fiancée had a miscarriage, experiencing the child stealer dimension of Lilith. For many years he had a significant relationship with actress Jane Asher and lived with her family for a few years. His constant infidelity, not unusual for Lilith in Gemini, eventually caused a break with her. He had a family with wife Linda, adopting her daughter from a previous marriage and having three children of their own. He has a daughter with now former wife Heather, who was not happy with her multi-million pound divorce settlement. He has a number of grandchildren.

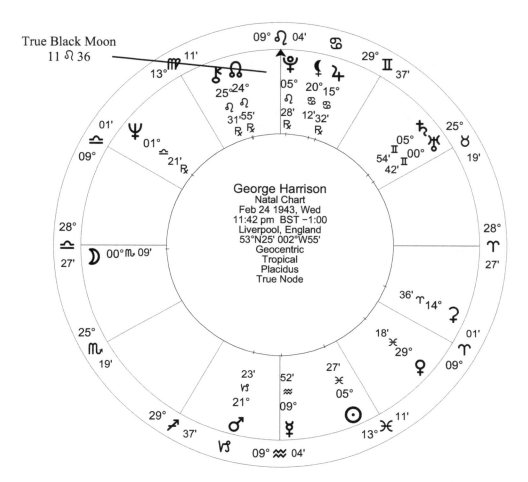

True Black Moon
11 ♌ 36

George Harrison
Natal Chart
Feb 24 1943, Wed
11:42 pm BST −1:00
Liverpool, England
53°N25' 002°W55'
Geocentric
Tropical
Placidus
True Node

George Harrison (mean Black Moon, 21 Cancer;
true Black Moon, 12 Leo; 21 degree corridor)

The "quiet" Beatle was born with mean Black Moon conjunction Jupiter in Cancer in the ninth house of travel, foreign connections and philosophy, reflecting his calling to the mysticism and music of a far-away place—India. He learned the sitar, studying with master Ravi Shankar, incorporating that instrument into the band's rich music. He was the group member most obviously involved with meditation, Hindu philosophy and the Hare Krishna tradition, playing a significant role in expanding awareness of these teachings in the West. Mars in Capricorn in the third house opposition Jupiter-Black Moon demonstrates the mastery of his business acumen; he became a successful music and film producer. The Mars aspects may also account for his interest in sports car racing.

His songs were less represented on Beatles albums than those of Lennon-McCartney. "Here Comes the Sun" is a favorite. "Something," written for first wife Pattie Boyd, pulls the romantic heart strings, and "While My Guitar Gently Weeps" augured his later spiritually-focused work. After the Beatles broke up, he released a triple album in 1970 called "All Things Must Pass" with such devotional songs as "My Sweet Lord," inspiring many on their spiritual journey. In 1971 he and Ravi Shankar organized Concert for Bangladesh, the first such rock music charity event.

George's 1980 autobiography, *I, Me, Mine,* fulfilled the publication potential of the ninth house placement. When first wife, Pattie Boyd, divorced George and married his good friend, Eric Clapton, the trio remained close in the unconventional lifestyle of the generation that came of age in the 1960s. Harrison's second wife, Olivia, with whom he had his son Dhani was Mexican-born. In 1988 he formed a band with fellow master musicians Bob Dylan, Roy Orbison, Tom Petty, Jeff Lynne with the ninth house name of Traveling Wilburys.

Whereas Paul has the Gemini-Cancer Black Moons, the Black Moon corridor in George's chart crosses from Cancer to Leo. True Black Moon in Leo conjunction Pluto on his Midheaven and opposition Mercury in Aquarius, all loosely square his Moon at the cusp of Scorpio, suggest extremism, so typical in the rock world, as well as the immense inner processing and creativity that helped carry his music and message into the world in a way that reverberated with the collective consciousness of his generation and beyond. A fine musician, Harrison made a unique and enduring contribution to expand the scope of popular rock and world music, but communicated so much more through his spirituality, compassion and quiet leadership. George died from a recurrence of throat cancer in 2001.

Ringo Starr (mean Black Moon, 3 Aries;
true Black Moon, 14 Pisces; 19 degree corridor)
A Cancer with Pisces rising, Ringo's true Black Moon in the twelfth house at 14 Pisces, is trine his Sun, while mean Black Moon, intercepted in Aries in the first house, is trine his Moon in Leo. These Black Moon sign placements are the same as Lennon's, who was born three months later, but are embodied with different aspects. The interesting interplay of water and fire in Ringo's chart reflects the retiring sensitivity and vulnerability mirrored in his famous Ringo hang-dog eyes as well as his lively personality, beguiling dead-pan humor and the energy he puts out behind the drum set. Interception designates an energy qualified by other signs and factors, indicating an urgency for fulfillment of some previously denied energy. An intercepted planet does not have as much "elbow room" and must find a special way to express itself. A first house that contains three signs gives a complex personality, and the Lilithian aura exudes a mysterious appeal.

The Aries Black Moon is trine a very tight stellium of Pluto, Mars, Moon and Mercury, packed with power and intensity. Perhaps "Don't Pass Me By" is a statement from that intercepted

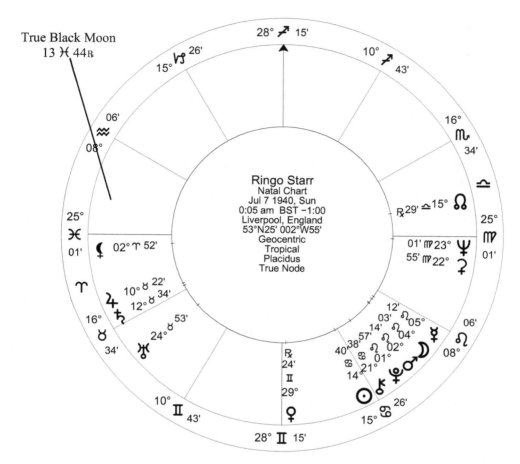

True Black Moon
13 ♓ 44ʀ

Ringo Starr
Natal Chart
Jul 7 1940, Sun
0:05 am BST −1:00
Liverpool, England
53°N25' 002°W55'
Geocentric
Tropical
Placidus
True Node

Aries Black Moon. Whatever that meant to Ringo personally, a surge of original creative expression was the result, with outlets through drums, songwriting and acting. It was destiny at work that Ringo even became a Beatle, brought in to replace another drummer. He was already well known in his own right. He developed special drum techniques and has a signature quality in his drumming that is said to be hard to duplicate. He was the central character in both the Beatles' movies, *Hard Day's Night* and *Yellow Submarine*.

His revolving member touring group, Ringo Starr and His All-Starr Band, features Ringo and Beatles classics but also showcases the talents of each musician, so each one is a star in true Leonine fashion. The intercepted Black Moon in Aries allows him to step back, but not out. Rather than the "lone star," he's the king pin. Ringo's musical collegiality works well for all. He has waved off talk of knighthood, which Paul has received. Outspoken against the monarchy, he quipped, "There goes me knighthood." So we won't be hearing of Sir Ringo. Although, that does have a ring to it, doesn't it?

How does the elusive true Black Moon in Pisces show up? Ringo's whimsical song "Octopus's Garden" (one of my favorites) musically interprets the water trine between his Cancer Sun and retiring true Black Moon in Pisces: "I'd like to be under the sea in an octopus's garden in the shade." He was the acknowledged behind-the-scenes hub of the Beatles and continues to maintain close friendships with former Beatles members and all the ex-wives and widows in a watery, feeling-connected way. He sobered up from years of alcoholism, eats vegetarian, and meditates on and off. He had three children with his first wife, Maureen. His oldest son, Zak Starkey, has also become a drummer, and has played in Ringo's All-Starr Band. His second wife is actress Barbara Bach, once a Bond girl in the 007 movie series. Ringo is a grandfather.

Beatle Asteroids

Asteroids 4147, 4148, 4149 and 4150 are named for the individual band members, while Asteroid 8749 honors what became more than the sum of the four: the Beatles. I just had to look up the Black Moon placements in the discovery charts of all five asteroids. Up-to-date data on all asteroids and some comets is available on NASA's Jet Propulsion Laboratory Small-Body Database Browser (//ssd.jpl.nasa.gov/sbdb.cgi.). Although no times are given, the dates, places and discoverers are documented along with the meaning of the name.

The most stunning revelation of my search was related to the discovery of Asteroid Beatles on April 3, 1998. The mean Black Moon at 12 Libra was in close opposition to the Sun at 13 Aries—the reverse of John Lennon's Sun-Black Moon opposition on his Nodal axis. That is a confirming "wow" to asteroid astrology!

It so happens I wrote this Beatles section on the weekend of April 3-5, 2009. Friday, April 3 was the eleventh anniversary of the discovery of 8749 Beatles, and the Beatles were in the news! Paul and Ringo were interviewed regarding the following night's charity concert called Change Begins Within held by the David Lynch Foundation to educate youth in the practice of Transcendental Meditation. The Beatles learned meditation in the 1960s from Maharishi Mahesh Yogi. Their spiritual retreat in India made an indelible mark on all of the Beatles, with meditation providing a stabilizing effect, Paul said in an April 3 interview. At the April 4 concert, Paul, with a little help from Ringo, played a roster of Beatles' and *apres*-Beatles songs, including his tribute to John, "Here Today."

On the same day, John's widow Yoko Ono unveiled her seven-foot mural called "Promise," showing layers of cloud on a blue field. This piece will be cut up into jigsaw pieces and auctioned as donations to autism research. George was in the news as well, via an announcement that he was to be posthumously honored with a star on the Hollywood Walk of Fame. His widow Olivia and son Dhani accepted the award on April 14 at 11:30 a.m., when there was a significant Venus conjunction within a minute of George's own Venus in the last degree of Pisces. How lovely. There was also a new release of George's music.

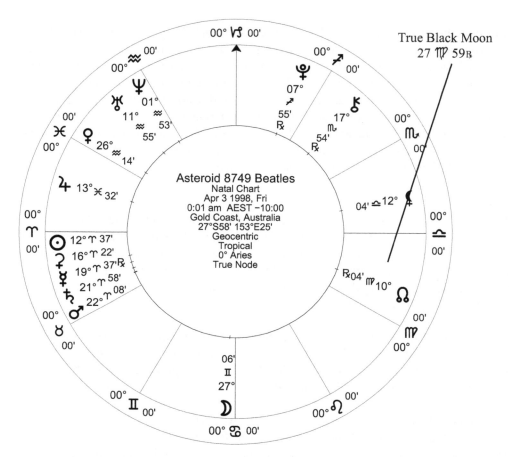

The strength of Lilith in the charts of the Beatles, individually and together, demonstrates something about the Black Moon's interweaving with the *anima mundi*, the world soul. These young men rose to a destiny that was greater than themselves, tuning into an unfolding wavelength in Earth's cosmically intelligent energy field as they played freely and courageously with the creative license of their time to forward our collective consciousness.

Black Moon can bring us so intimately in touch with the light body of Earth that its cosmic creativity pours through us, whether we are consciously aware of it or not, and certainly more easily when mental interference is short-circuited by that "Something" of Lilith:

> *Something in the way she moves . . .*
> *something in the way she knows . . .*
> *something in the things she shows me.*

Chapter Four

Black Moon in the Houses

From ancient times, Lilith has lived in the tree of life. But where does she live in your chart? In what house does the Black Moon Lilith dwell?

The astrological houses are like the stage sets of your life. The presence of the Black Moon in any house by birth will add inscrutable layers of complexity to the area of your life represented by that house. By transit, the house through which Lilith is passing may be where the most compelling challenges pull your energy, yet also where special potentials beckon. Black Moon invisibly lights up the chakras to a huge influx of energy which needs to be integrated or discharged in some way.

The transparency of the Black Moon clean-sweeps all shadowy corners, wipes clean the dusty, musty, rusty, untended, funky, behind-the-toilet, under the sink areas. We can't blame the clean rag for the dirt. Black Moon exposes false imprinting and offers transcendence. Potential pathways of expression open up, as something in us desires liberation of soul and spirit. Thus the Black Moon moves through the houses. Difficulties often arise from our interference, largely unconscious, due to our misunderstanding about what really is going on. Indulgence in self-doubt, self-satisfaction or any other self-oriented attitudes will short-circuit fulfillment in the house Black Moon occupies, as Lilith has no patience for such a waste of time. At its best the Black Moon vortex, void of substance, cuts through the extraneous to express unusual clarity and rare qualities or abilities.

Often the presence of the Black Moon is played out by a particular person or a type of person represented by the house placement, typically someone who pushes significant buttons. Friction with this person comes from an underlying tension in deep karmic layers, either with this individual or from your own psychological scars with persons in similar situations. The more you can objectify the relationship and dis-identify from the role you play in the push-and-pull of direct engagement, the more the manipulative undertones ease. Often this is a lifetime process, but in its unfolding you can find some value, at least for your soul's growth.

Of course the zodiac sign of the Black Moon placement colors its expression and is integral to the interpretation. Because of the particularity of the Black Moon, the combination of sign and house needs to be carefully fused. Signs and houses are two different things. The tendency to conflate the meaning and houses and signs leads to confusion and distortion, especially in such specific interpretations as the Black Moon requires. Most astrologers learned that Aries and the first house are similar, but with Black Moon, take care to discern the quality of energy from its realm of application.

The following interpretations by house are just starting points to consider as you begin to intuit the unique implications of the Black Moon influence in a chart. Again, I want to emphasize that Black Moon resists cookbook interpretation regarding her expression. My wish in offering these suggestions is to stimulate your own astro-intelligence and to encourage you to trust your own guidance, asking as many questions as you need to ferret out the subtext influences of the Black Moon through addressing Dark Moon, a different astronomical Lilith, I have found the astute insights in Delphine Jay's *Interpreting Lilith* useful to Black Moon investigation. Examples of famous persons only touch on the depths of Lilith for we cannot know the inner experience of that person. Perhaps the public fascination with these persons stems from the mysteries of Lilith, evoking a deep response.

I had my own experience with the Black Moon transiting my sixth house while investigating and concentrating to articulate this interpretational approach to Black Moon houses. When I got to houses nine (my natal Black Moon location) and ten, it so happened that a visiting friend helped me rearrange my home office area in the large main room of my apartment. We applied the principles of feng shui. The new spatial arrangement still calls for some fine tuning, yet has already evoked a response from my internal space in relation to my daily work flow. Why was I surprised that new writings about Lilith in a concentrated way would have such an effect? The process was as intriguing as the theoretical material, for the Black Moon is imminently embodied.

First House

In this house of self-projection, *you* are Lilith in the very real way that the transparent Black Moon energy impacts your imprint on the world. Whether you know it or not, the deep dark of the Black Moon exudes through your presence. Sooner or later you will become aware of the potency of the energy you project by the responses of other people as they react to you in ways that make you wonder who they are looking at. You are often a mystery to yourself as well as others even if you have been aware and perhaps highly self-conscious about your appearance and persona since your early life. Many Lilith people recognize some heightened dynamism or distinction that sets them apart.

The impact of the Black Moon is of course stronger the closer it is to the Ascendant, or if it is conjunct a planet. Some Black Moon personalities revel in their charisma and magnetism; others are seduced by it themselves. The Black Moon is not a factor you can claim for yourself. If you do, it will lead you into precarious places and get you in trouble. Lilith is an ego-buster and requires that you get out of your own way. You can be the vehicle for an extraordinary light to enter the world and touch the hearts of others if you willingly give yourself to it.

In chapter three, there are many examples of people with Black Moon on the Ascendant, demonstrating their particular Black Moon uniqueness. One prime example is that of Norma Jean Mortenson, who projected a potent Black Moon charisma. She could turn on her iconic Marilyn Monroe persona at will, yet it was also the agent of her undoing.

One needs a strong sense of self to be able to carry the Black Moon without letting it undermine. In one sense, when an individual carries such a powerful archetype in the persona it can overshadow the personality in a way that makes it very difficult to function outside of the often exaggerated archetype. Yet the Black Moon can become the dancing *dakini*, the guide of the soul's journey to fullness of being. Monroe was a fearless explorer of the art of acting, quite serious about developing her full potential well beyond the sex symbol image. Black Moon can be helpful in exploring the different versions and aspects of one's own living mythology, engaging the evolution of consciousness offered by the subtle layers of the Black Moon.

Any character in a myth is more fully known through its interactions with other players in the story. As in a dreamscape, all characters come back to the mirror the Self. Life is but a dream itself. The Black Moon points a laser light toward this perception, urging a deep movement of the soul to awaken from the blindness of this dream. With Black Moon on the Ascendant, one is an agent of this awakening for others as well, and in no superficial way.

A short roster of some who carry this Black Moon house placement are John Lennon, Gene Simmons (of Kiss), Tom Cruise, Shakira, Picasso, Beyonce, Joan Baez, Krishnamurti, Sharon Tate, Johnny Cash, David Lynch.

Second House

Matter and physicality are explored in this sector of the chart. This is a banking and accounting house that contains resources and assets. Often thought of as primarily money and finances, various other resources are included, such as talents, an old word for money. Black Moon in this house can indicate the capacity to make a lot of money, yet it becomes very obvious that money can be difficult to handle. It can be the need for money that compels individuals to dig deep and tap into unused abilities they never suspected they had and to develop rusty skills into fine-honed assets.

Outer assets and the way they are used reflect the value you place upon yourself and your abilities, or may be used as a substitute to prove worthiness. Issues about money, often related to the sense of self-worth or unworthiness, lead to a drive for a great deal of money or on the other hand a tendency to be underpaid, charging too little for one's services with the result of chronic financial short fall.

Your body is your primary possession and physical vehicle, your temple of the spirit. The way you care for this primary asset demonstrates your level of self-esteem and how you honor or dishonor much about the relationship of the soul to the material world, to matter. If we take one "T" from matter, we have mater, mother. The Great Mother gives us all there is. Lilith as the Great Mother in this sense pays attention to the bottom line of karmic implications that ensnare the soul in matter. The Black Moon often exaggerates a subliminal self-abnegation that undermines healthy self-esteem and financial well-being. A type of greed can be present that seeks to fill a hungry hole. Intense desires for objects and possessiveness can lead to disaster in Lilith's highly impersonal presence. Big gains can be followed by huge losses to test how closely you identify with your possessions.

Famed general and self-proclaimed emperor Napoleon Bonaparte conquered an empire, power and wealth through brilliant military strategy, accompanied by plunder, extortion and other such tactics, described in part by true Black Moon in Scorpio. This placement in the first house is followed by mean Black Moon in the second in Sagittarius square a Neptune-Mars conjunction in Virgo. Black Moon in aspect to Mars indicates an intense drive and potential to push the edge of any limits. He lost power and empire in extravagant folly, overestimating his ability and perhaps his luck. He came back from his first exile to claim power once more and lost it yet again at the famous Battle of Waterloo. His charisma carries on through the centuries.

Astronomer Galileo Galilei was capable of technical precision that allowed him to discern the workings of nature and sky. This father of astronomy had an active Black Moon in Virgo conjunct Sun-Venus and square Neptune. His conflict with the church that led to his house arrest in late life was fueled as much by his belligerent "I-told-you-so" attitude regarding his paradigm-shifting scientific discoveries.

From rags to riches, Walt Disney built the world's largest media empire with its own Magic Kingdom, not unlike the Vatican. His shrewd Scorpio Black Moon-North Node conjunction urged him to develop his abilities and take them ever further, sometimes at great financial risk. He embraced the latest technologies to bring his inventive ideas into new forms. He was known for obsessive-compulsive habits and hobbies, and his legacies are the many sub-personalities portrayed in his beloved cartoon characters.

Third House

In this sector of siblings, perception, learning and communication, Lilith may wear the mask of a brother or sister. Issues with sibling rivalry are common. It may be the death of a sibling before you were even born or the lack of one that haunts you, or the illness, disability or other disturbance of a brother or sister that focuses family attention.

In the third house of learning and perception, the Black Moon may stress early learning and education, whether through learning disorders, visual or auditory anomalies, some type of unique schooling or another related issue. There may be something extraordinary in one's way of thinking or the ways and means of communication that can feel like a blessing and a curse. Perception may be finely tuned to certain interests, channeled into a specific tunnel vision that can be at variance with others around you. Although this can be a special gift, it also can be a challenge to accept it as such or figure out how to communicate it. We don't know her astrology, but Cassandra, the cursed Greek priestess and seeress of Troy, would likely have had this placement. Her prophetic visions were disregarded, leading to tragedy.

Icelandic singer, song-writer and actress Bjork has a wild Black Moon in Aquarius. A triple Scorpio with Neptune on the Ascendant, one or the other Black Moon squared all Scorpio factors for a pronounced Lilith flavor, impacted by the calculated abandon of experimental Aquarius. Bjork is known for her eccentric, quirky style of techno-electric music that appeals to the same sort of fans. She thinks in music, she says. Winner of the Best Actress award at the Cannes Film Festival for her tragic role in *Dancer in the Dark*, she also wrote a song for that movie and was nominated for an Oscar. A more Plutonian or Lilithian movie you cannot find, with odd, unexpected interludes of song at key moments that express the character's fantasy life. She is highly respected for her artistic concepts and integrity.

The chart of Jackie O, more formally known as Jacqueline Bouvier Kennedy Onassis, has mean Black Moon in Capricorn in the second house, demonstrating her privileged background, elegance, fine taste and support of the arts and historic preservation. But her true Black Moon was most actively involved in her chart, placed in early Aquarius in the third house opposition Sun-Mercury. Her only sibling was socialite Princess Lee Radziwill, who married a Polish prince. The media loved Jackie, for better and worse, in good times and bad. Yet some of Jackie's greatest satisfaction came from her editorial career pursued quietly later in life.

Belgian tennis star Justine Henin's one-handed backhand, one of the best in the game, may be a gift of her Sagittarius Black Moon in the third house. Closely conjunct Neptune, this Lilith may also account for her award as the first Whirlpool Sixth Sense Player of the Year for the intuitive intelligence of her play. Yet another Gemini tennis pro (we find a full roster of them in this quintessential Gemini sport), her true Black Moon is in the second house, even stronger than the true, conjunct Uranus and opposite Sun-retrograde Mercury. Perhaps it is this house placement that urges her to take her skills as far as she can. She has left the game and come back twice as of 2010, perhaps torn between her Sagittarian desire for independence and furthering her goals in the world of sports. One key to success is her mental toughness on court, a sign of a strong third house. Roger Federer also has this trait and this Black Moon house position.

Fourth House

Family matters are inevitably complex and painful to some degree. One parent in particular, probably the one of the opposite sex, plays Lilith. There is an endless variety of scripts for the family dynamics Lilith can create here. Somewhere in the family ancestry there is a hidden nugget of gold, but it may take some sorting and sifting through muddy waters to find it. Once found and freed, you can move deeper in to the archaeology of your own soul history to unearth layers more akin to your spiritual roots.

An uncommon sensitivity to what goes on inside and under the surface of the Earth confounds the individual with this placement until it is identified for what it is: an energetic plumb line to connect heaven and Earth. On the journey of life, one seeks a true home and place on this globe which is ultimately beyond these planetary shores and into other dimensions. The way there may include dislocation from homeland, exile or abandonment, and yet success in a foreign country. It is typical to feel uprooted, like a stranger in a strange land looking for one's soul tribe.

The person with this placement needs a living situation in which she or he feels totally at ease. The principles of feng shui, the oriental art of placement for clear energy flow, may appeal and be applied. Those sharing this person's space need to be easy to be around and comfortable; otherwise, the Lilith person may be constantly anxious about taking care of them. The Lilith person may be happier in his or her own space, preferably a stand-alone dwelling. It may be essential to at least have "a room of one's own."

Virginia Woolf, author of the essay of that same title, has Black Moon in Leo square a complex stellium of Saturn, Neptune, Jupiter, Chiron and Pluto in Taurus. Her mean Black Moon is in the fourth house opposition Mercury. The true Black Moon is a few degrees back in the third house opposition her Aquarian Sun in the ninth. Since the Sun rules the fourth house, this opposition brings attention back to the Lilith influence in that house. Her mother died when she was thirteen, just as she was heading into her teens, and then her father died when she was in her early twenties. She had eight siblings and half-siblings (third house). Early sexual abuse by two brothers likely contributed to lifelong sexual ambivalence. She also had a stormy and probably physical relationship with one sister. Without parents, this literal house of siblings became a literary salon of the *avant-garde*, the Bloomsbury Group. With her husband Leonard she started a printing business (third) in their home (fourth house). They published their prolific works, fiction and social commentary, as well as the work of other major authors, including T.S.Eliot, the Freuds, Rilke. The fourth house represents the end of life. She took her own by drowning, seeking to end her continual suffering from emotional neuroses and nervous breakdowns.

Another woman writer mentioned earlier, Anais Nin, has Lilith in this same house. Like Woolf, she wrote fiction and voluminous diaries. She later relocated to the U.S., place of her professional success. Her complex family life included adult incest with her father and marriage to two men at the same time, one on each U.S. coast.

Archibald Leach, better known as Cary Grant grew up thinking that his mother was dead, and only as an adult did he learn that she was alive and had been institutionalized by his father. He left home at sixteen, relocated to the U.S. and performed on stilts. The change of citizenship became permanent when he became a U.S. citizen, largely to avoid being called up in WWII. His five marriages gave him a stepson who died young and a daughter when he was sixty-two, over whom there were custody battles after divorce.

Charles Manson was the central "father" figure of his "family," forming a tragic dependency that led members to commit murder. Manson himself was a deeply wounded child.

Michael Moore is a different example of a fourth house Lilith. As a man of his native land, he is a citizen activist who wants it to live up to its potential.

Fifth House
This is the sector of children, creativity, love affairs, recreational play, leisure, theater, games and sports. In short, this is where you do what you do because you enjoy doing it.

Lilith in her role as child killer may take her toll in this house of children, resulting in death or lack of children, and perhaps a child with a handicap or one with unusual brilliance who may require special attention. That child plays the part of Lilith. One mother has challenges from her children. One daughter, who worked as a pole dancer, was convicted and imprisoned for drug use. Her son is mentally disabled and unable to be self-sustaining. The mother herself has a high creative output focused on therapeutic work with groups and families.

In this house of love affairs, Lilith can become the playgirl or the playboy, maybe even a playboy bunny. We can "play" with this keyword, widening the usual scope of playboy beyond sexual dalliance to mean, truly, playing. Parties, social events, games and sports are the playgrounds of these jet-setters, leisure rich or layabouts. Like Peter Pan, some men don't want to grown up. They'd rather play, be it on boats, in sports or games. Some "girls just want to have fun"—all the time. Creative expression may serve as a necessary outlet for emotional intensity and lead to an original oeuvre. Sometimes people with Lilith in this position are capable of extraordinary excellence through hobbies.

Bill Gates became a multi-billionaire by playing around with computers, a hobby turned into a business. The true and mean Black Moons in Sagittarius sandwich his North Node, one on either side of the sixth house cusp. He has to love his work, and indeed, even his vacations are think-tank sessions. His creative intelligence was inspired by a vision of the future potentials in computer technology. The Bill and Melinda Gates Foundation contributes some of his billions to philanthropic work and education, mainly to support the well-being of the world's children.

Fellow philanthropist Bono shared *Time* magazine's 2005 Person of the Year award with the Gates' the same year his band U2 was inducted into the Rock and Roll Hall of Fame. He leads a double life, both as the wildly creative superstar lead singer of U2 and as a well-informed global activist productively hob-knobbing with world leaders in his inevitable jeans and shades. U2 has been a key contributor to major musical fundraisers. Bono co-founded DATA, an organization to raise awareness about Debt AIDS Trade Africa. He has two Black Moons that indicate the dual streams of his passions. True Black Moon in Gemini is another such indicator of multiple interests that somehow and sometimes controversially co-exist in some way particularly unique to this man. Just across the Cancer cusp, the mean Black Moon is more active in his chart it opposes his Jupiter and creates a grand trine in water signs with Chiron in Pisces and Moon-Neptune (and Dark Moon Lilith) in Scorpio.

Janis Joplin lived on the wild side during her short life as a female breakthrough artist in the rock and roll scene of the 1960s. The Black Moon in Cancer conjunct her Moon-Jupiter reflects her unhappy childhood and tendency toward excess. In the fifth house, she let loose any and all emotional hurt in her wailing whopper of a voice. A no-holds-barred play and party girl, she overdosed on heroin, her drug of choice, at age twenty-seven. Her progressed Moon had just returned and was two degrees from an exact conjunction with her Black Moon.

Rocco Siffredi lives out one fantasy of Black Moon in the fifth house as actor, director and producer of pornographic films. He has starred in more than 1,300 such films, pushing the edges of hardcore porn. Nicknamed "Italian Stallion," he is especially known for a long penis and somewhat deviant sexual specialties. His Black Moon is in Sagittarius, the true position widely opposite Venus in Gemini.

Other popular examples of the unique creative expression that Lilith can evoke in this sector include Prince, Bruce Lee, Cat Stevens, Anne Frank, Keanu Reeves, George Clooney, and Robert Downey, Jr.

Sixth House

In the house of work, health and service, Black Moon requires a level of humility that is not a natural trait of all signs. This is the arena of the work-a-day world. Lilith in the sixth needs to work according to one's own standards and rhythms. The workplace environment, both physically and socially, requires a quality described by the sign involved. Aries makes its own way, disliking authority. Taurus is most productive in a comfortable physical space or in the field. Gemini needs a lot of interaction, changes of pace and frequent breaks, whereas Cancer needs more intimate personal connections. Leo requires appreciation and freedom of expression. Virgo has a meaningful and particular skill set to use. Libra prefers to work with partners, and likes to choose them. Scorpio zeros in on the job at hand and dislikes interruptions or to be imposed upon. Sagittarius needs expansive horizons beyond an office cubicle. Ambitious Capricorn has a master plan and operates at the controls of the work station. Aquarius is a team player on equal terms. Pisces has its own flow and needs down time. Lilith is the ultimate authority, no matter who the boss may be.

Black Moon here is specifically rigorous about clearing the physical system to enhance the cohesion in the mind-body interface. Illness can result when the outer world activities are unrelated to inner experience. This creates tension, stress and dis-ease, often difficult to diagnose. Health issues can stem from deep waves of subconscious or superconscious energies unrecognized by the conscious self. Symptoms are highly specialized when pinpointed by Lilith, located in subtle body systems and detected and best treated by attention to those delicate layers. Neurotic tendencies can be present, sometimes related to what is going on globally or deep inside Earth. When one's work suits the authenticity of the Black Moon, it can be exquisite, award-winning and of great service, sometimes accompanied by personal sacrifice.

Mohandas Gandhi was a pioneer in service of his people with both Black Moons in Aries in this house opposite his Libra Sun. True Black Moon conjunct Neptune clearly indicates his spirituality, his struggles to live up to his own ideals and his work as spiritual service. True Black Moon conjunct Chiron is way outside the norm and suggests his unique dedication to simplicity, his courage to act, his basic idea to reclaim the harvesting of daily salt for the Indian people and, in the end, his martyrdom.

If the reported 7:15 a.m. birth time is correct, highly regarded actress Kate Winslet has mean Black Moon in Pisces square Mars in Gemini. Her feel for a range of unusual characters accounts for some of her best work. Her true Black Moon at 2 Aries forms a grand trine with Saturn in Leo and Neptune in Sagittarius. As an actress she is fearless and serious about the authenticity of her work, and had been nominated for six Academy Awards by age thirty-three.

As songstress and actress for more than four decades, Cher has won many awards in music, film and television in the U.S. and abroad (Sagittarius), including an Oscar for her role in *Moon-*

struck—how Lilith of her! Her Black Moons in Sagittarius straddle the sixth house cusp, with mean Black Moon in the fifth close to opposing her Sun at the very end of Taurus. True Black Moon in the sixth is opposite Uranus and is close enough to the South Node to feel a sense of destiny. Her #1 top song titles sound like keywords for her Lilith: "Gypsies, Tramps and Thieves," "Half Breed," "Dark Lady," and "Believe," this last particularly appropriate for the Sagittarius placement. With top-ten hits in five decades, Cher has become iconic on the global stage. She has set fashion trends and pushed censorship edges over the years.

Cher's fifth house children are both singers. Her daughter Chastity or Chaz (with Sonny Bono) was gender-challenged. Cher became an advocate for gay rights because of this, and thus her huge fan base includes a strong gay contingent. Chaz has now changed sex and become an activist for transgender persons. Cher's son Elijah (with Gregg Allman) is a vocalist and guitarist with Deadsy, an artsy band that fuses many esoteric and magical elements into their music. Cher's sixth house service orientation includes political and anti-war activism and major support for active U.S. troops as well as disabled veterans and other charitable groups.

Seventh House

The opposite of the first house, this sector features Lilith as played by significant others. A relationship with a Lilith type person does not include a cozy home surrounded by a white picket fence on Main Street, even if that is what it looks like from the outside. The partner is likely to be an accurate mirror of some of your deepest fears or resistances. Remembering that Lilith trips us up when the ego takes things personally, a person with a seventh house Lilith is called to impersonal experiences in a most personal area of life. Partnership can be the hallmark of the life, for better and worse, as the influence of the partner looms large. The partnership can take on a life of its own, leading one astray or to great heights.

The quality of the complex dynamics of Lilith at play in relationships is described by the sign it occupies. Some issues of contention that can apply with a seventh house Lilith through the signs can be gleaned by reading the section on Ascendants, as if that is the "other" in the relationship. Think of the best and the worst of each sign to get a brief take on what kind of issues can be expected in relational processes.

Aries has a willfulness and selfishness that tends to put self in front at all costs with no regard for the partner; or it can be such a clear example of self-integrity that it requires the other person find a finer level of authenticity. Taurus can be materially and sexually greedy and possessive, yet can support a partnership through extreme ups and downs.

Gemini can be accused of immaturity and whimsy. It is not a sign known for its constancy and is one of the most likely to be homo- or bi-sexual. Gemini is also one of the most tolerant and open-minded signs.

Cancer's defensiveness and tendency for dependency can feel suffocating to some, but it can give and nurture on a level more caring than most. Leo needs a lot of attention and can create a lot of drama, but no one can have more fun, appreciate a partner or be a better example of heart and courage. Virgo can be so nitpicky and critical that nothing can ever please, including the partner. At its best, Virgo can naturally attend to delicate nuances of relationship beyond the skill of any other sign.

Libra can crowd a partner, looking to the other for answers and decisions. On the other hand, Libra's capacity to truly listen to another can be a win-win for a true partnership. Scorpio's demanding nature can take manipulation to an extreme, bringing out the worst but also the very best in another person.

The overbearing nature of Sagittarius can be self-righteous in its certainty of knowing what is best for everybody else. They are often right and hold the light for others to see. Capricorn can approach a relationship with a controlling, brusque business approach. It is a sign that tends to mature like fine wine and over time can grow a relationship into a successful partnership. As cold and detached as Aquarius can sometimes be, eccentric and freedom-loving, the offer of space and freedom of individuality is available if you are willing to live up to it. Shapeshifting Pisces can play the sacrificial victim better than any other sign, but the quality of compassion can lift up love to the sweetest realms of soul-to-soul communion.

Lady Diana Spencer had Black Moon in the house of relationship. Her first marriage was marred by deceit from the beginning because Prince Charles had a lover. The fairy-tale wedding was a mirage before it even took place. The relationship with her second fiancé, rich Egyptian playboy Dodi Al-Fayed, was doomed for many reasons, opposed by the powerful British royals, if not his Arabic family as well. Religious background was an issue, shown by his Black Moon in Sagittarius. They perished together in a car crash.

Influential philosopher Frederich Nietzsche demonstrated in a very specific way the joint themes of the sixth and seventh houses. True Black Moon Lilith stands in the sixth in Taurus with mean Black Moon in the seventh, quite strong in Gemini opposite Moon in the philosophy sign Sagittarius and square Venus in Virgo. How much this intense feminine dynamic influenced his views is worth a study. The Moon's Nodes fell on his mean Black Moon (South Node) and Moon (North Node). Brilliant, yet with fragile health, he obtained a professorship in his early twenties. By his midforties he exhibited serious mental illness and cared for by his sister (Venus) and mother (Moon) for the rest of his life.

John Wayne Gacy was a sexual predator and serial killer, murdering more than thirty young men and boys. Married twice, his second wife divorced him on the grounds of impotency. Both Black Moons in Gemini were significantly active in his chart and suggest his double life and knack of engaging others, like a salesman. The mean Black Moon was conjunct Mars-Jupiter, suggesting exaggerated yet ambivalent sexual needs. The true Black Moon is square his Sun-Moon in Pisces in the fourth house and Neptune in Virgo in the tenth. A sensitive Pisces, he must have suffered greatly from his dysfunctional family headed by an abusive alcoholic father. While in prison pursuing a protracted appeal process, he enjoyed his notoriety, setting up a 900 number and selling his paintings through the mail.

Barack Obama's First Lady is his Lilith. With Liliths in Leo in the sixth and seventh houses, his partner is his queen and equal helpmate in his life work. Lilith in this house also reflects his relationship to the public, on the world's main stage. He can put on a good show but doesn't take it for granted. On a life mission, he does the work.

Eighth House

The Black Moon thrives on its own terms in this house so often associated with sex, death and taxes, as well as power and desire. These are all areas that are intimately connected with others or with other realms. Natives of this placement feel subtle nuances and karmic roots in the interplay and energetic interdynamics with other people and other worlds. Such persons have a heightened, often painfully exquisite awareness of the effect of influences, impositions, psychic and psychological undercurrents from others. On a worldly level, this house plays out in the various ways we transfer energy with others, sexually, financially, psychically. There are very strong threads connecting these areas. Lilith will do some of her finest work weaving these strands into sophisticated dynamics.

The eighth house covers the whole territory of subconscious and unconscious motivations and the impact of our desires, as well as the impact others will have upon us. Black Moon is resistant to inappropriate inroads and will maneuver the energetic membrane to increase the level of psychic immunity.

Controlling persons, often early authority figures, can invade one's energetic space. One result of this is the person seeks situations and spaces to be free of the influence of other people. Some strategies and pathways to this self-integrated space are healthier than others and some are illusory; Lilith insists that we overcome judgment, and the edges are always tested. Since we are all connected energetically, there really is no place to get away, and ultimately we are drawn to explore inner space, from which can emerge a universe. Once a center point is found, it can release a huge amount of regenerative energy that carries a powerful transformative effect. Healing, the empowerment of others and the impact on the collective field are potential results. This search for the ultimate often leads to compelling engagement with the esoteric realms. Black Moon Lilith offers a capacity to plunge into the hidden mysteries if you are willing to take her up on her dare. If you don't, she'll find a way to make you want to go there.

Sexuality is a major theme of this house and we can expect Lilith to go to extremes, seeking to release the self into something more than itself. Orgasm is likened to a kind of death. We surrender and feel amazingly fluid and free. But it is not ultimate; the experience needs to be repeated over and over for the ecstasy, as the search or true bliss continues. Death is one ultimate release, so Lilith in this house can evoke a fascination with death, that of self and/or others. Death-related experiences can be awakeners or leave deep scars. Lilith in this house can bring extremes of agony and ecstasy.

One woman with this placement in Taurus suffered through two increasingly difficult stepmothers.The first was religiously intolerant, the second one took her inheritance. She is a very private person and a counselor-teacher. Another woman has Black Moons pinpointed in the same degree in Sagittarius. She is a genius at gathering people together for spiritual exploration, ritual and mutual empowerment. She has been a part of several large groups in healing and esoteric research and education, and has traveled the globe on spiritual journeys with a close group of friends. A third example is that of a women with Cancer Black Moon placements in this house. Family obligations pull at her in uncomfortable ways, creating deep resentment. Some deep strand of inherited obligation and connectivity compels her to engage with the clan. I would guess there is some financial push and pull involved.

Tiger Woods gave us a scandalous example of Lilith in Aries spanning the seventh and eighth houses. His eighth house placement of true Black Moon is conjunct Chiron trine his Moon in Sagittarius. The mean Black Moon in the seventh is opposite Pluto. Jupiter is at the midpoint of the two Black Moons, adding an extra sparkle to his talent and the potential for an inflated pride. In the seventh this can show the high esteem placed in him by the public as well as the bigger fall from such a high pedestal. In addition to putting his marriage into jeopardy, another result of his exposed exploits and sexual addiction has been the withdrawal of lucrative corporate endorsements, resulting in huge financial loss and reflecting the interactive financial dynamics of this house. Woods had one of the highest incomes of any sports super star. Many public discussions have ensued about his impact on the sport of golf, both because of his excellence and the scandal. What kind of behavior do we expect of our iconic public figures? Apparently Tiger both exceeded and failed public expectations.

Frank Sinatra had Black Moons in Gemini in the eighth house opposite his Sagittarius Sun-Mercury in the second house and square ruler Jupiter in Pisces in the fifth house. "Ol' Blue Eyes" had a stellar career with his clear velvet voice and also became an Academy Award-winning actor. He made huge amounts of money from his talent and investments. Associated with Las Vegas, he is known for having Mafia connections and cavorting on the "dark side." He associated with power players and was one himself. He was also a generous philanthropist.

Ninth House

Lilith in this house of higher mind, higher education, international travel and universal law is likely to wrestle with angels in defiance of organized religions that codify Truth into defined belief systems. Black Moon here has an eclectic philosophy and seeks to push back the horizons of knowledge. These persons look for core truths of human existence that underlie all religions and philosophies, making significant discoveries along the way. Esoteric and foreign territories are explored in a thirsty quest for such truths; otherwise, this person may be the ultimate agnostic.

One truth-seeking woman finds herself immersed in philosophical and spiritual traditions from many cultures. She contends with deep questions about how to synthesize and fuse threads from the various paths she explores along her journey.

The legal field may appeal or be forced upon one in some way, becoming a likely background to further interest in the higher or cosmic laws. Publishing is another realm of ninth house activity through which Lilith may insist upon being expressed, as I have found through my own placement of the Black Moon here.

International travel tends to play a significant part in life, or at least connections with persons from other countries. Lilith here is the exotic stranger. Ninth house Lilith persons may play an influential role on the global stage, in politics as ambassadors or power players, such as Margaret Thatcher and Condoleezza Rice.

Deepak Chopra is a good example of one who has taken his native philosophical tradition, penetrated its core with a broad range of studies and refashioned those teachings for the modern mind. Indian-born, he has found success in a country foreign from his birth place. The Black Moon placement in Sagittarius on the South Node continues a philosophical search of many lifetimes, ready to be articulated afresh and published for an international audience.

Oprah Winfrey has also made millions as she educates a hungry public on a broad range of topics not previously addressed on television. Her Black Moon in Libra suggests the one-on-one interview format she uses with great success. Black Moon creates a grand trine in air signs, with its trine to the second house Venus, Sun and Mercury in Aquarius, indicating her wealth, and Jupiter in Gemini in the sixth house. Her background of personal health and weight issues gives her a special interest in healing that she shares with others. The ninth house position of the Black Moon is also demonstrated in her African philanthropic activities, magazine publication and book club. She has also been involved in movies.

One of the most esteemed actresses of her time, Meryl Streep has created an astonishing roster of shining performances that have earned more Oscar nominations than any other actress. The

Black Moons in Aries are square her exact Sun-Uranus conjunction. This combination shows how she pushes herself to reach into the essence of any character, and the trine with her Leo Ascendant helps her find a way to dramatically express it. The ninth house placement contributes to her special ability to take on accents from many different countries. In 1978, when the Black Moon entered Cancer, it crossed her Sun-Uranus and then Venus, a life-changing year. When a close male friend died, she sublet an apartment from the unknown friend of a friend, soon met this unknown man and married him.

Tenth House

This house is associated with the individual's career, standing and reputation in the world. A career that allows one to channel the intensity of Lilith will bring out the individual's best, extraordinary talent capable of outstanding brilliance and success. Success is not the ultimate goal, however; rather, Black Moon's agenda will be fulfilled by making a unique contribution. This may be off the beaten path and possibly unappreciated for a long time, even beyond the lifetime. Out in the world, the person will represent Lilith, often at her best and worst. Perhaps the individual will embody a quality that has little to do with his or her personality, or the person can be obsessed to express something greater than oneself.

Most often a Black Moon Lilith native of this house of status and reputation is and feels over-exposed to the world. Black Moon insists upon being out there, so the person can't hide or disguise. If a person does not find his or her true professional calling, a deep frustration will create problems, a sense of failure, inadequacy or some disappointment that leads to a career crisis. Parental impact, particularly of the authority figure or lack thereof, can generate issues that interfere with successful career development. Many lucky people find their calling early on, even if it is not considered a career option until later in life.

Often such people want to be in charge of their own career and feel chaffed by the imposition of bosses and "higher-ups." Bosses can play the Lilith role in lieu of the authoritative parent. Lilith holds no one higher than herself, so there can be a king-of-the-mountain tension with people in authority. When the level of personality self-consciousness infiltrates the career stage, Lilith "shows her teeth." Indulgence in self-doubt, self-congratulations and other self-centered emotions will short-circuit fulfillment in this house. Lilith has no patience for such activity, which distracts from the work at hand. A sense of serving something higher than oneself brings out the best of what the individual has to offer.

U.S. President John F. Kennedy had less than a degree separating his two Leo Black Moon positions in the tenth house. He was a kingly leader, famous for his Leonine-titled book, *Profiles in Courage*, which won the Pulitzer Prize in 1957. When inaugurated in 1961, he became the youngest and first Roman Catholic president.

His Black Moons were in a volatile T-square: opposition Uranus in Aquarius and square a Taurus stellium of Mars, Mercury and Jupiter in the eighth house. We have touched on the classic connections of the eighth house to death, money and sexuality; Kennedy had them all. Born to a very wealthy family, he was a rampant womanizer, a fact little known until after his death. Father of two children, he and wife Jacqueline also experienced a miscarriage, a still-born child and one who lived for less than a week. His own assassination shocked the world and remains suspiciously secreted.

The Black Moons in his inauguration chart were in Cancer, the mean conjunct his Saturn on the Midheaven, the true conjunct his South Node in the ninth house. Uranus was conjunct his Black Moons and opposite his natal Uranus, setting off the T-square. His world-beloved First Lady, Jacqueline Bouvier Kennedy, had a close Ascendant-South Node in Scorpio that filled in the empty corner of JFK's T-square. Her mean Black Moon in Capricorn was conjunct his North Node, and her true Black Moon in early Aquarius was opposite her Sun-Mercury conjunction on his tenth house Neptune. She was queen to his king, partnering his (illusory) legacy of a lost Camelot.

In Hollywood we find numerous examples of Black Moon in the tenth, including such luminaries as Julia Roberts, Renee Zellweger, Audrey Hepburn, Jim Carrey, Drew Barrymore, and Jodie Foster. Two-time Oscar winner Foster has the Black Moons just three degrees apart in Libra in the tenth. A very private actress, perhaps we see her Black Moon expressed most clearly through her life-long work, as actress, director and producer. Her first Academy Award for best actress came from her role as a rape victim in *The Accused*; and the second from playing a murder investigator in *The Silence of the Lambs*, co-starring with Anthony Hopkins as the cannibal Hannibal Lector. In the movie *Contact,* she went on a psychic journey to the star Vega. She was harassed by a stalker, John Hinckley, Jr., who shot President Ronald Reagan to claim her attention.

We find two scientific paradigm shifters with Black Moon in the tenth—Dr. Albert Einstein and Nicolaus Copernicus. Copernicus, father of modern astronomy, has a mean Black Moon-Gemini Midheaven opposite Jupiter in Sagittarius. True Black Moon was twelve degrees farther into the tenth in a T-square, closely conjunction Saturn and square Sun-Mercury in Pisces and Pluto on the Virgo Ascendant. He proved for the modern era that the Sun is the center of the solar system, creating the "Copernican Revolution."

Centuries later, Einstein's famous equation, $E=MC^2$, described the fusion of energy and matter, inaugurating the quantum era and a breakthrough in theoretical physics. His discovery of the underlying photoelectric effect as an essential cosmic field earned him the 1922 Nobel Prize. His true Black Moon was at two degrees of the trail-blazing sign of Aries in the tenth house, heightening the intelligence of a Mercury-Saturn conjunction. An odd boy, he was late to speak and considered slow, possibly mentally retarded. No doubt he was wired differently, plugged into a subtle aetheric Lilith energy system. The titles of his first three papers, published in 1905, tell us much about Lilith's subtle realm of activity: "On the Motion of Small Particles Suspended in a Stationary Liquid According to the Molecular Kinetic Theory of Induction"; On a Heuristic Viewpoint Concerning the Production and Transformation of Light"; and "On the Electrodynamics of Moving Bodies."

When the mean Black Moon crossed into Einstein's eleventh house, a rebellious streak against the "establishment" led him to quit school as a teen and to speak out against Hitler's Germany. He became a U.S. citizen, known for his social activism as a pacifist and sponsor of German-Jewish refugees. Involved with the Princeton'Institute for Advanced Study, Einstein continued his research outside mainstream physics. At the end of the twentieth century, *Time* magazine named him "Person of the Century." Einstein has entered the collective consciousness as something greater than his personality; he has become an iconic embodiment of genius.

Eleventh House

In the social realm, the eleventh house experience connects us with circles of people, friends and organizations and evokes our response to trends and ideas of our times. Those with the Black Moon in this sector influence or are influenced by the interface between inner calling and the larger social context of their lives. Even more, they interface with the larger tapestry of history and even pre-history, with a radar-like scanning capacity that transcends time and space.

These people have a heightened awareness, consciously or unconsciously, of the energetic interplay in social situations. They can be "in the world but not of it," or in a group but not really a wholehearted part of it, on the outside looking in. One can even feel like a pariah, the black (moon) sheep. Such a situation may be Lilith's message to look elsewhere. The Black Moon seeks a deeper field of connectivity than one finds at the mall or through social media, although there are already hundreds of Lilith references on social media sites and thousands on the internet! Lilith has a huge presence in the digital airwaves. As the opposite of the fifth house, Lilith in the eleventh best applies personal creative energy to a socially relevant theme.

One woman was highly self-conscious in group situations and tended to stand out as different, or felt that she did, She could discern unspoken feedback from others. As she explored various forms of original creative expression, particularly within a group context, she became adept at allowing the creativity of others to come forth. She holds the space of a group as a powerful cauldron of deep personal alchemy.

As always, the sign placement largely qualifies the way to seek these true (or mean!) networks, where one can contribute most effectively. As a group member, this one can be the life of the party, or a true leader that fosters the quality of the team. There is the possibility of taking on the "shade" of a group or being the most excellent spokesperson, as long as the vision is in synch with Black Moon's desire; otherwise Lilith won't be responsible for the results. One can act out the lowest common denominator of the group dynamic.

Black Moon here has a deep capacity for friendship and for collegiality, although it may last only as long as the common purpose and vision remains without personality issues getting in the way. Lilith can wear the face of a friend, a coworker or a fellow group member. Any ego trips are mirrored back in this fun house of mirrors. Friendships can change over time as Lilith thins the veils. Betrayal or a completion of some kind may bring an abrupt end to certain relationships. It may not be possible to continue individual relationships outside of the community context in which they were created.

As Black Moon seeks to set the soul free, it looks to associate with others doing the same, not necessarily on the very same path, but perhaps so. However, Black Moon is not cultish and if any group association leans in that direction, problems arise and experience can turn negative, even scandalous. Black Moon Lilith stands for transparency. Perhaps we most honor these persons for the transparency of some greater social idea coming through them, not always for the personality example.

Bhagwan Shree Rajneesh can be considered an example of the best and the worst of this position. Yet who can judge Lilith or an enlightened being? Rajneesh embraced multiple streams of philosophy, psychology and spirituality from both East and West in his teachings. A man of high intensity and compelling charisma, he was able to brilliantly express levels of understanding and consciousness. He drew many followers to a path of tantric yoga and eventually created an ashram in Oregon, intended to be a model of ecological and spiritual living. Huge spending and difficult relationships within the ashram leadership as well as the local community and government eventually led to the arrest of Rajneesh and his deportation back to India. Many in the community had committed their lives to his work and teachings and were forced to recreate their lives upon the dissolution of the ashram.

His Black Moon positions in Aries pushed the edge and demanded a personal clarity and integrity that can be hard to maintain in a power position, especially when beholden to others for financial support. His teachings were radical and so was he. Mean Black Moon was very close to Uranus in Aries, retrograde, and the hub of the T-square in the early 1930s with Pluto in Cancer and Saturn in Capricorn. Rajneesh had a strong Capricorn energy with Mars, Mercury, Venus and Moon in that sign. Venus and Moon are in the eighth house, most closely square Black Moon. With those feminine factors in a dynamic interaction, the goddess enters in full force, denying any patriarchal approach. Any Capricorn control issues in the sexually and financially active eighth house would have been undermined by Lilith.

Another excellent example of an eleventh house Black Moon connected to esoteric organizations is Alice A. Bailey, who was a well-known esoteric writer and astrologer. She founded the Arcane School that has influenced many astrologers since its beginning in 1921. She and her second husband, Foster Bailey, organized the International Goodwill Movement, which is linked with the Arcane School. She is especially known for her series of books cowritten with "the Tibetan," an ascended master, an unusual colleague with whom to collaborate. Her true Black Moon in Cancer is on her South Node, conjunct Mercury and square her Moon in Libra, altogether suggesting a stream of consciousness carried over from a the past. Indeed she wrote dozens of books on ancient teachings. Her mean Black Moon is strong as well, in Gemini, conjunct her Sun and Venus, the esoteric ruler of that sign.

In the music world, we find Sting of the hugely popular band The Police, who then went on to his own successful solo career. His true Black Moon in Gemini trine Neptune-Moon was deeply responsive to creative musical interactions of just the right group, and thus able to hit a unique note on the social scale. Sting's mean Black Moon is conjunct Uranus and square Sun-Neptune-Moon, just tucking into the twelfth house for some solo work. He sings of sacred love, is interested in yoga and higher consciousness.

Leonard Cohen's Black Moons span most of Leo and the eleventh house, the Black Moon being most active in conjunction with his South Node-Mars and square to Uranus in Taurus. Other indications such as an exact Neptune-Venus aspect in Virgo in the twelfth square Chiron in Gemini in the ninth support his unique, moody poetry as well as his calling to Zen Buddhism. The emotional angst of Lilith in Leo, seeking free expression, picked up a key tone of his times, as he exemplified some of the issues of his generation. Though he retired in the early 1990s to a monastery, the world has not forgotten him. A tribute album, "Tower of Song," came out in 1995, followed ten years later by the documentary *I'm Your Man*.

President Franklin D. Roosevelt and First Lady Eleanor Roosevelt both had Black Moons in the eleventh house. FDR had the Black Moon in Leo, the leadership sign, opposition Sun, Venus and Mercury in Aquarius and square a potent Taurus stellium of Saturn, Neptune, Jupiter and Chiron in the eighth house. Assuming the presidency in the middle of the Great Depression and through most of WWII, FDR held on into a historic fourth term before he died in office. He was in much pain from a polio-like disease for years. He is remembered particularly for the economic initiative called the New Deal, creating significant social programs.

Eleanor Roosevelt, a powerhouse in her own right, had true Black Moon in Scorpio conjunct ruling planet Mars, both in a close trine with her Cancer Moon. Her Scorpio conjunction filled in the empty corner of FDR's T-square. If this scenario is sounding eerily familiar, look back at the eighth house vignette on Jack and Jackie Kennedy, who had planetary patterns similar as these earlier White House residents. As potent a position as it is, the Scorpio Black Moon does not tell the whole story in Eleanor's chart, as mean Black Moon is found on her Sagittarius Ascendant. She had just about the widest Lilith corridor possible, and could not be contained to the U.S. alone. She followed her own social agenda, befriending groups and leaders and becoming the hub of an intellectual elite. Well known on the international stage, she has been dubbed First Lady of the World.

Twelfth House

This house of mystery is a natural haunt of Lilith, though she'll make herself at home anywhere. This house holds deep secrets and skeletons in the closet. It can be considered a kind of karmic backpack carried throughout life. Black Moon here is very sensitive, open to psychic energy and currents of the collective consciousness.

As the house of retreat and seclusion, the Black Moon can compel a reclusive need to withdraw from the impact of the outer world. Worldly activity can distract and skew the clarity of reception to signals coming from inner space. Whether through writing, art, ideas or meditation and contemplation, this person will seek communion with realms beyond this one. The need for privacy can be urgent, beyond conscious recognition or personal volition. It can sometimes be hard to know if the solitude is mystic or neurotic escapism: Black Moon defies such judgments. In the twelfth house, the depth of experience cannot be understood by worldly outsiders.

Withdrawal from active engagement in the world does not eradicate psychic participation in the collective consciousness; indeed, Black Moon in this house can enhance sensitivity to the aetheric currents of the times. Some natives of this placement are born in sheltered or remote circumstances in such a way that they are not in sync with modern times. Some feel a sense of exile. Some are fragile, physically, mentally or emotionally, in a way that keeps them apart from others, pained by contact in some way. One can't fight Black Moon. A surrender to circumstance can allow the flow of its true impulse to come through, then may follow an ineffable upliftment.

Actress Greta Garbo was famous for her declaration of privacy, "I want to be left alone." She had mean Black Moon in Aries, the true in Taurus. She often worked in solitude even on a movie set, surrounded by curtains that invited in only the camera. When alone, she could explore deep inner spaces that evoke rare nuances that penetrate emotionally beyond the movie screen. Her post-WWII return to the silver screen was short-circuited. She lived in relative solitude the rest of her life.

Auguste Renoir, a leading member of the Impressionist movement of the nineteenth and twentieth centuries, had both Black Moons in Capricorn. The true position in the twelfth is strong, square Pluto-Moon-Venus in Aries. He helped to invent the new artistic vision through his extraordinary sensitivity to the luminous feminine divine. He left an *oeuvre* that includes a number of the most popular paintings in the world. Mean Black Moon in the eleventh shows his involvement with the Impressionist group of artists who co-created this historic artistic style. His mastery of the form is indicated by the Capricorn placement.

Called "Queen of Crime," Agatha Christie is one of the greatest mystery writers and best selling authors of all time. She could weave intricate plots with great psychological sophistication and developed two detectives that became as famous as herself: Hercule Poirot and Miss Marple. Her play, *The Mouse Trap*, is the longest running in London theater history. Her Black Moon is in Leo, the sign of drama. It forms a quintile (known as the "genius" aspect) with the Pluto-Neptune conjunction of her decade, located in her tenth house.

Famous Swedish director Ingmar Bergman had Black Moons two degrees apart in Libra, conjunct Mars (true BM exactly) and square Pluto. He spoke of his demons, with whom he wrestled in his dark and moody films with stark imagery and psychologically-nuanced characters. The title of his film *Through a Glass Darkly* evokes the Black Moon. He wrote his own scripts, drawing from his early religious questioning and existential sense of loneliness. He gave actors great leeway in exploring and expressing their inner landscapes in his films, as if something greater than his own visions were being chanelled through the medium. The eleventh house placement of his Libra Moon, still close to the Black Moons, suggests his sense of co-creative collegiality. He established a long-term professional partnership with his cinematographer and worked with a company of actors who appeared in several films. He left Sweden for several years in a self-exile while being investigated for tax fraud. In twelfth house fashion, he spent his last years, again in self-exile, on the Swedish island of Faro.

Reclusive poet Emily Dickinson stayed in her father's house and garden, never venturing more than a few miles from home. "The Soul that has a Guest rarely goes abroad," she wrote. Her principle dwelling was her inner world, where she tended the life of the soul and wrote short poetic verse tied in bundles and hidden away until after her death. Much of this contemplative verse considered Life, Death, God, Love and Eternity, all of which haunted her life in mystical reverence. With Black Moon in Scorpio, the true position very close to her Ascendant, she was intently appreciative of the miracle of life and of Nature. "Alas, that wisdom is so large and truth so manifold!" she exclaimed in ink. "The only news I know is bulletins all day from Immortality."

Chapter Five

Black Moon Lilith with the Planets

O, swear not by the Moon, th'inconstant Moon
That monthly changes in her circled orb,
Let that thy love prove likewise variable.
—Juliet to Romeo in Shakespeare's *Romeo and Juliet*

Following some initial impressions and global examples of Lilith by sign and house place-ments, we can begin to consider the Black Moon in relation to the planets. The closer the better with aspects. Some Black Moon researchers use only conjunctions and oppositions, easier as-pects to spot in action. Conjunctions are obviously prime. Again, Lilith is subtle and gets so per-sonal that much depends upon how she calls to you, what you notice and respond to.

Lilith is not a planet, so looking at Lilith in relation to planets requires a different orientation. The Black Moon represents a vital mathematical factor in the gravitational dynamic between Moon-Earth, influenced by the Sun, so perhaps it is more like an eclipse point. Earlier it was mentioned that the Black Moon's orbit is a half-cycle of that of the Moon's Nodes, which mark eclipse points. Certainly Black Moon brings a darkening of the light within and beneath our daily consciousness to evoke the light in the dark. An eclipse reveals a face of the dark goddess. During a solar eclipse the disk of the New Moon, normally invisible, passes across the face of the Sun. In this relatively rare moment, we see the dark disk of the New Moon, the silhouette of the dark mother. During a lunar eclipse the Moon is "turned off" and darkened or reddened. Still visible, the Moon's face changes color, reflecting the Sun's light in an altered light. Perhaps in those moments the Black Moon is more turned on and more expressive in its ever-present invisibility.

Like an eclipse, a Lilith passage is life-altering or life-altaring, potentially made holy in a black hole-type effect. The Black Moon lifts her veil to sacred encounter, while exposing all the shadows of our apparent unworthiness, a humbling discovery of the true self as both human and divine. Black Moon resonates in the aetheric realm, aether being the subtle element of consciousness that underlies and gives rise to the four tangible elements of life that we work with in astrology—earth, air, fire and water. As we tune into Black Moon we are interconnecting with All-That-Is. Black Moon is going to show up the obscured sectors where our vision is clouded. Ouch! Those are the areas that cause pain and rage. Our reaction can be one of striking out or clutching in and pulling back, holding on for dear life. We need to reconsider (con = with, sider = stars) what we hold as "dear" in life, real life. If life truly is but a dream, Lilith may enhance that dream in her illusory play until we grow out of the game. We come out from under clouds that obscure the light. As the movie musical title suggests, "On a Clear Day You Can See Forever."

The aetheric body is intertwined with our physical bodies through the endocrine system, spinal nerves, chakra centers and other biological systems. This subtle body channels energy, prana, life force through all body systems. Perhaps the Black Moon could be said to rule this aetheric body, the cosmic "star body." The high side of Lilith is the soul's desire to be released into this free space, unfettered with karmic necessity that must, nevertheless, be paid out and played out. The Black Moon puts us through the paces with hidden intent, difficult to trust. At times she faces us with what doesn't seem reasonable or even possible.

In search of your Lilith storyline, hold in mind the mythic her-story of Lilith as seductress, child killer, sexual initiator, primordial mother and lover, magical woman, mystic muse. Lilith-loving astrologer Spencer Grendahl offers evocative interpretational phrases articulating her polarities such as "door to the spiral of mystery, door of pain between the dimensions, magnetic, mesmerizing, repulsive, revolting, giver of VD, home wrecker, husband stalker, cradle rocker, cradle robber. terminator/initiator, yin from outer space, fertility deeper than the sea, knower of the darkness over the deep, solar wind messages from the invisible to the eye" (personal communication).

Lilith resonates with other dark goddesses or witches of secret mysteries and creative power. Clearly associated with magic, she ultimately regenerates us through direct experiences of sacrifice, surrender, sexuality and mysticism. Obsessive behavior may be witnessed with any planet impacted by the Black Moon. Consider her themes of autonomy, exile, extremism, and denial. Black Moon experiences have a deeply internal effect, tuning into subtle signals of soul stimulation. Such moments are likely to occur along with troubling ripple effects in the mundane world, as the mirror of life reflects the imminence of soul awakening and the dissolution of an illusory aspect of our personal soap opera. In Lilith's territory we are confronted with some of our most difficult karma (past life leftovers and obligations of this life time) and soul contracts with ourselves and others, but also with the potentially sublime.

The horoscope is like a cosmic play. Each planet is an actor, costumed according to sign, on a stage set represented by its house placement. The aspects are the dialogues between the actors, be they ally or enemy, in cooperative or challenging geometries. Lilith is not in the original cast. She is like the thirteenth fairy not invited to the party for fear of what she might do or say. Of course she says and does it. If we open to her *wyrd*, the unique loops and twists of our unknown destiny, in pivotal moments of our lives that reveal the soul, then she may open the path less taken, most truly desired.

The question of Lilith's intention has haunted human history and imagination. As the serpent, what is she offering Eve and Adam? The answer to this question has engaged artists for thousands of years. What does Lilith mean to us in our time as we stand at the threshold of a new age, one created from deep co-creative engagement with the Cosmos? We each come to this quest. We have already tasted the Apple. We are here. How do we open the smallness of our personal being to sacred encounter? Lilith leads the way. Or does she? We each discover our own relationship with Lilith and with our Selves. As the Dark Matter, Mater or Mother of our inner divine child, she feeds us what most nurtures our soul in its growth. Referring back to Juliet's plea above, Black Moon seeks the invariable light of love.

The Personal Planets: Mercury, Venus and Mars

Mercury

This quicksilver planet of perception translates experience into thought, developing interpretations based on data. Like its mythic namesake, Mercury has a trickster side, with its monkey-mind antics. The Black Moon will seek to unravel rationality and offer another level of perception: Mercurial curiosity peeking into dark edges, the Mercury-Black Moon duo can pick up on ideas that are "out there" in the aethers, well beyond the radar of most people. Mental wiring can be connected to a rarefied power source, coming from a cellular connection with a remote cosmic broadband.

Black Moon can be mentally diamond-sharp, and Mercury ready to say "The Word." As perception becomes increasingly transparent, less clouded by personal agendas, cultural overlays and influences from the past, one's interior world becomes more vibrant. At the same time it can become less possible to communicate such perceptions in normal language. Silence can be the

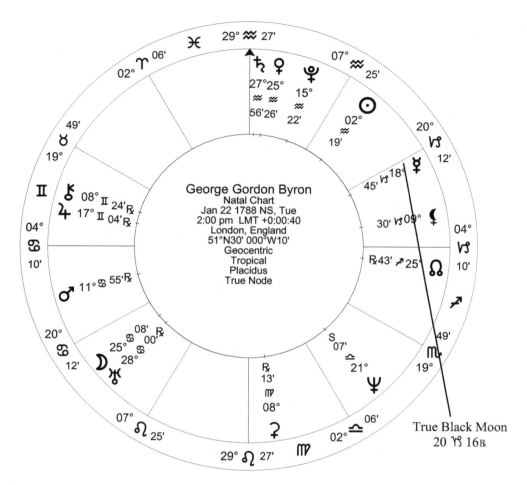

George Gordon Byron
Natal Chart
Jan 22 1788 NS, Tue
2:00 pm LMT +0:00:40
London, England
51°N30' 000°W10'
Geocentric
Tropical
Placidus
True Node

True Black Moon
20 ♑ 16ʀ

most eloquent. Music and poetry or symbolic imagery become truer languages of the soul, especially when Venus is anywhere close. Still, what is communicated may be but a mere shadow of the truth, which may appear paradoxical or even downright falsehood.

We saw that Paul McCartney has a Black Moon-Sun conjunction in Gemini, which is joined by Mercury as ruling planet. He gave musical voice to more than one generation. Barack Obama was also mentioned earlier, with his true Black Moon conjunction Mercury to the degree. Agnetha Faltskog was a leading voice of ABBA, the Scandinavian pop music group that gave us *Mama Mia*. With her Mercury so close to the Black Moon in Aries, she also sang solo with Lilith. Listen to the words.

The writings of George Gordon, Lord Byron, epitomized romantic melancholy and guilt, accompanied by a yearning for nobility of spirit—"half dust, half deity, alike unfit to sink or soar"—in a style that continued to influence later generations of writers. His Black Moon-Mercury conjunction is in a vital T-square, square Neptune in Libra and opposite a Moon-Uranus conjunction in Cancer, with Mars retrograde in Cancer pulled in. His writing captured the spirit of the age and of ages to come. With such a sensitive Moon in Cancer in that mix, inevitably his childhood was unsettled and rather isolated, with a mentally unstable mother and a club foot. Byron came into his great-uncle's fortune and title early on, but his life was spent largely in exile, unconventional living and adventuring. His two children died. He himself died at age thirty-six while supporting the Greek resistance against Turkey. Interestingly, his same Uranus and Neptune placements came around again in the 1950s, during a a modern social revolution that rebelled and yearned. If you were born between 1952 and 1956, the writings of this poet may speak to you.

My friend Lila was born in 1994, in the special generation with the Uranus-Neptune conjunction in Capricorn. Lila has her mean Black Moon and Mercury at 3 Libra. She has had deep dreams from childhood, using dream catchers at an early age. She has a special rapport with animals. She loves drawing them and has made an animal calendar with her photos. Birds would land on her hand when she was quite small. She is already showing talent as a writer. What will she need to learn to fulfill her longing? Lilith will show her. I look forward to hearing what she has to say.

Venus and Mars

Placed on either side of our planet, Venus faces us inward toward the Sun, Mars outward toward the rest of the solar system and beyond. These mythic lovers are the Earthling's emotional yin and yang. The subtle energy of Black Moon underlying either of these planets can bring emotional/sexual experiences of heightened intensity. The laser beam of Lilith's desire power inevitably requires us to let go of attachments, especially obsessive ones. With Black Moon connected to Venus and/or Mars, relationships can be agony and bliss because surrender to the bottomless well of love can be simultaneously painful and ecstatic. As our emotional bodies lighten, we begin to feel and generate light, a cornucopia of unconditional love. The betrayals on the way may be of ourselves.

Venus

> . . . *a so forgotten beauty that she couldn't even exist as a remote memory;*
> *Lilith, the immortal Queen of all moonlit nights, the Queen of the Lilin!*—Thorp
> McClusky, "The Evocation" (quoted on //lilith.abroadplanet.com/Images.php)

Like the Moon, Venus is a developmental aspect of feminine experience in both women and men. Venus is the magnetic attracting energy that vibrates with an emotional resonance based on our self-esteem and sense of self-worth. This planet applies the Law of Attraction. Black Moon pulls one into deep engagement with this theme, in a drastic or more kindly way depending on the intensity of our emotional self-love or self-rejection. Attractions are pulled in from very deep for soul satisfaction, which we consciously know little about. Some of these attractions come in order to finish "old business" so we can move on.

This combination stirs up questions about values and aesthetics. Natural paths of expression are through relationships and through the arts. Relationships influenced by Lilith are not promising for a typical kind of happiness, but are soul contracts with deep intentions. Half-hearted efforts are not allowed. The emotional and sexual charge can be felt in a compelling, sometimes compulsive way, disguising deeper issues. The obsessive aspects of Lilith can contribute to emotional implosions or remoteness in relationships, depending on sign. The ultimate aim is for re-creativity and the transformational potential of the surrender to love.

"Love reveals us to ourselves," wrote Jungian analyst Aldo Carotenuto in *Eros and Pathos: Shades of Love and Suffering*. This daunting potential is a motivation of Lilith-Venus interactions and of Lilith-infused relationships. "Love so activates the possibilities of self-knowledge that an entire universe heretofore unknown offers itself to the wide and curious eyes of the lover," continued Carotenuto, "a spark of the infinite." Yet, as the old saying goes, "Love brings up everything it is not." Negative emotions are inevitably present along with the positive. Do we dare bare our soul to the beloved, another flawed human? Perhaps this combination spurs us on to relationship with the divine within, passionately expressed in devotional literature, such as the Biblical Song of Songs, the Psalms, the poetry of Rumi and Omar Khayyam (with Black Moon, Sun, Venus and Mercury, both retrograde, in Gemini).

The sparkling screen presence of Elizabeth Taylor shows her Venus-Lilith in Aries. She is mentioned later in the Uranus section as well. We hear Lilith's voice in the poetry of R.M. Rilke, whose Black Moon-Venus conjunction is at 1 Capricorn. Was it Venusian Lilith that pricked him with the rose thorn that caused his death? Alfred Hitchcock's Black Moon-Venus conjunction may account for the dark nuances, psychological complexities and the emotional thrills and chills of his films.

A popular recording artist known for his high stage theatrics (Pluto in Leo rising), Peter Gabriel has the Black Moon in Taurus at the Midheaven square Venus in Aquarius. His discography expresses the storyline of his Venusian Black Moon in a way that melds the personal and socially relevant. As lead singer with Genesis, he wrote the lyrics for "Lilywhite Lilith" on the album, *The Lamb Lies Down on Broadway*. Give it a listen. She'll "take you through the tunnel of night." A leader in the world music scene, he also created The Hub on YouTube for human rights awareness. He is one of the Founders that support The Elders, an international group of recognized leaders who create initiatives to address global issues of pressing concern.

The famous 1969 Woodstock Music and Art Fair: An Aquarian Exposition had the Venus-Black Moon conjunction in Cancer, close to the U.S. Sun. This event is remembered as a

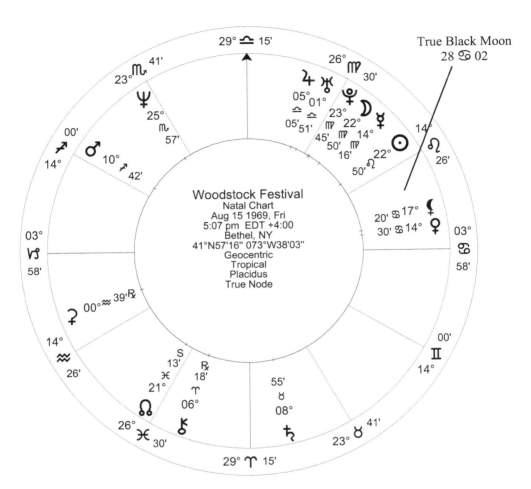

tribal gathering of a half million people from across the country, creating a deeper sense of bonding in the counterculture community. The social harmony was remarkable for such a large event. Beautifully reflecting this Cancer placement, the event took place on a dairy farm. The Pluto-Moon-South Node conjunction in Virgo helped make it the huge collective event it was. The date may have been chosen by an astrologer; this was the era of *Hair* and the words from this musical: ". . . dawning of the Age of Aquarius." I was there. Were you?

Mars

Black Moon-Mars persons often have fascinations or confrontations with sexuality, with violence or with the yang assertion in life. Black Moon can turn things inside out; in association with Mars output, Lilith can pack a punch and can even redefine what true strength is. Lilith can instigate obsessive behavior until her message comes through and the energy is opened to another level of expression. The "wants" and "NOT wants" can both be strongly projected. Men

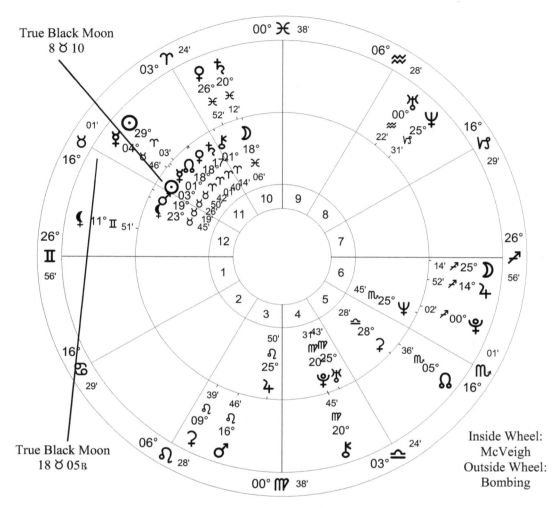

True Black Moon
8 ♉ 10

True Black Moon
18 ♉ 05ʀ

Inside Wheel:
McVeigh
Outside Wheel:
Bombing

redefine maleness; women must claim their Martian autonomy in relationships, with men and women.

We met a few Black Moon-Mars examples in earlier sections: Suzanne Valadon, HRH Josephine-Charlotte and Madeleine Albright all had Mars conjunct the Black Moon. Albright walked the patriarchal corridors of power, holding her own. The lives of these first two women, both with Black Moons in Libra, were largely shaped by the men in their lives. Surrounded by famous male artists, including husband and son, Valadon pursued her own creative passion and has endured as an artist in her own right. The royal marriage of Josephine-Charlotte defined her life. Nelson Mandela makes an interesting addition to this list, with Black Moon conjunct Mars in Libra and square Jupiter-Pluto in Cancer. Confronted with apartheid violence and imprisonment, he found a way beyond to a political solution and more peaceful coexistence. The movie *Invictus* is based on the true story of Mandela's active support of South Africa's rugby team (Mars = sports) to unify the country. D.H. Lawrence wrote groundbreaking literature about sexual relations between men and women, still censored by some. His mean Black Moon in Capricorn is opposite Mars in Cancer and square Venus-Moon in Libra.

Timothy McVeigh, the Oklahoma bomber, had a fascination with guns and violence. His Sun-Mercury, and especially Mars, all in Taurus were infiltrated by the two Black Moon positions. The mean Black Moon-Mars is closely square Jupiter in Leo and opposite Neptune in Scorpio. A Gulf War veteran, awarded with a Bronze Star, he had a hard time adjusting to civilian life. He engineered the bombing of the Alfred P. Murray Federal Building in Oklahoma City on April 19, 1995, killing 168 people and injuring hundreds more.

At the time the bomb exploded, mean Black Moon was on the Gemini Ascendant (many young children in a day care were taken by Lilith the child killer), while true Black Moon was on McVeigh's Mars-Black Moon. The chart of the event also has Mercury-South Node on his Sun-Black Moon. "Isn't it kind of scary that one man could wreak this kind of hell," he asked in a *New York Times* interview. He was executed in June 2001. At that moment, the Black Moon was conjunct transiting Uranus and the Moon on McVeigh's Midheaven. He wanted his death publicly aired on television, but that request was denied. The death chart Sun took him in his twelfth house, and Venus in Taurus was exactly conjunct his Sun-Black Moon and the Mercury-South Node of the bombing. Jupiter, Sun and Mercury were on his Ascendant opposite Chiron-Mars on his Descendant. The execution gave to some a sense of karmic satisfaction.

The Social Planets: Jupiter and Saturn

Jupiter

King of the Gods meets Lilith and a great mythic plot unfolds. Black Moon opens Jupiter beyond its normal broad scope, challenging core beliefs beyond cultural and religious boundaries. Jupiter's "bigger than life" effect can amplify the Black Moon's agenda, broadcasting its meaning into the social fabric and enlarging the scope of possibility. There is a tendency to overdo in the area of life indicated by this conjunction.

As the Big Planet is exalted in Cancer, we can expect to find some notable examples of Jupiter-Black Moon in Cancer. Indeed, we find several interesting clusters of famous persons. Jupiter is in a sign for a year, hanging out with the Black Moon on and off while in the same sign. Born under the 1907 conjunction we find iconic movie stars Katherine Hepburn, Sir Lawrence Olivier and John Wayne, who merit mention in the Neptune section as well with a triple conjunction of Jupiter-Neptune-Black Moon. Jupiter was conjunct the Black Moon again in 1942-1943 to oversee the births of Jimi Hendrix, George Harrison, Janis Joplin, Bobby Fischer,

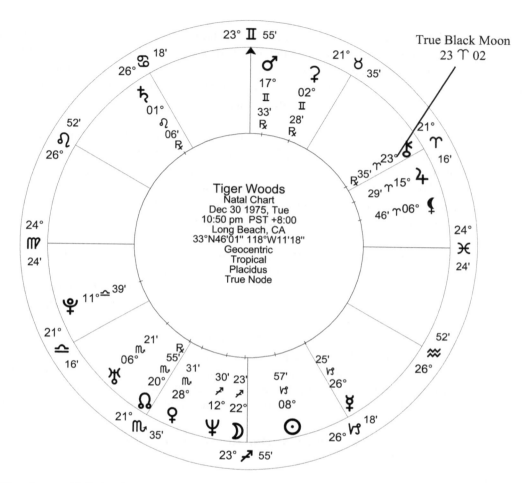

True Black Moon
23 ♈ 02

Tiger Woods
Natal Chart
Dec 30 1975, Tue
10:50 pm PST +8:00
Long Beach, CA
33°N46'01" 118°W11'18"
Geocentric
Tropical
Placidus
True Node

Ben Kingsley and Michael Palin (Monty Python), all of whom made outstanding contributions to their fields and to the culture. Sharon Tate is in this group as well, a tragic example that expresses the Lilith child-killer face in Cancer, sign of motherhood. George Harrison's conjunction is in the ninth house, emphasizing Jupiter's search for philosophical meaning in devotional Cancer.

Jupiter-Black Moon in trail-blazing Aries manifested in the sports arena through a pair of 1975-1976 super stars that achieved record-breaking success: Peyton Manning in football and Tiger Woods in golf. Tiger Woods is a notable Aries original. A golf prodigy, he was putting with Bob Hope on the *Mike Douglas Show* at age two and featured in *Golf Digest* at age five. He has a unique heritage that he calls "Cablinasian," a mix of Chinese, Thai, African American, Native American and Dutch. He is Buddhist, as is his mother. As a man of color breaking almost every record, he challenged the elitism of the standard golf world and repopularized the sport.

True Black Moon is in the same degree as Chiron, closely trine his sportsy Sagittarius Moon and square Mercury. Recent scandals reveal his obsession with sexuality, suggested by true Black Moon in the eighth house conjunct Chiron.

Saturn

Delphine Jay, writing about the Dark (or Waldemath) Moon Lilith, notes similarities between Liltih and Saturn, the stern planet. This similarity carries over to the Black Moon, both with their tough edges. Black Moon seeks embodiment as much as Saturn does; both of these factors are thus inherently manifestational, but the means to the end differ. Black Moon works from outside the known world, and Saturn from within it. The two together can redefine parameters and paradigms, exposing, brushing aside and pushing past the boundaries of the status quo. Issues with the personal father, the internalized or religious "father" or the patriarchal worldview can be contentious. The authority is necessarily and authentically inner-determined. Like most Saturn-related experiences, the finest expression of mastery tends to come later in life. Even if there is great success early in life, the true path unfolds later, standing up to the test of time. even "fathering" new conventions.

Noam Chomsky is a good example. Black Moon is conjunct Sun-Saturn in Sagittarius, indicating his broad range of intuitive insight into the workings of the human mind and tell-it-like-it-is political activism. Andy Warhol, pop artist of the modern era, is another Sagittarius example with South Node, true Black Moon and Saturn in a multiple grand trine with Uranus-Moon in Aries and Mercury-Sun in Leo. His soup can is one of his famous statements, or questions, about the relationship of pop culture and art. Is this his Saturn-Black Moon showing how the mundane transcends itself?

Another artist with this conjunction, this time in Capricorn, ruled by Saturn, was Auguste Renoir. He chronicled his age in once revolutionary, now-classic Impressionist paintings, infusing his work with his strong Pisces flavor. Vincent Van Gogh also had this conjunction in Taurus; he was yet another artist who took his form to new level.

John Ronald Reuel Tolkien, professor, philologist and self-described "hobbit," was a mythmaker extraordinaire, creating a whole world of "Middle Earth." Born on an apex of the Neptune-Pluto conjunction in Gemini that occurred in 1891-92, he helped usher in this new cycle. His work continues to stir the collective imagination. Saturn and Black Moon, both well-placed in Libra, are trine that outer planet conjunction and square his Mercury-Sun in Capricorn. The Saturn factor was large in his life. He lost his father when very young and his mother at age twelve. Earth sign Capricorn can be the most in touch with nature kingdoms and the world of Faerie, with an unexpected quirkiness to offset the seriousness they experience in life. *The Hobbit*, a playful invention for his children, was a huge success, when it came out in 1937. His

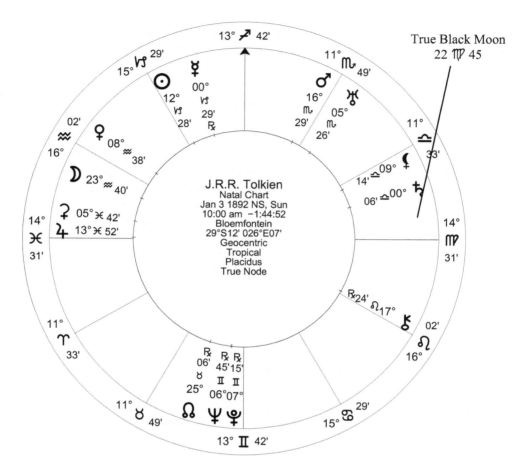

Saturn-Black Moon made a solid relationship with a unique dimension of the other world. His brilliance in philology, understanding the structure of language, led him to create his own languages. Using Finnish and Welsh, he created a language he imagined would be spoken by elves. He labored over *The Lord of the Rings* trilogy for a decade, creating a unique "virtual" world, archaic and archetypal, relevant for today yet anti-modern, a heroic adventure story with emotional depth and enduring values. Oddly old-fashioned, he refused to authorize a paperback version, so sales were slow until it was pirated, republished and properly publicized. In a time when myth was considered fantasy, Tolkien firmly felt that myth and fairy tales have meaning inherent to the human experience: "The imagined beings have their inside on the outside; they are visible souls. And Man as a whole, Man pitted against the Universe, have we seen him at all till we see that he is like a hero in a fairy tale?" (from Tolkienlibrary.com)

Richard I, the Lion-Hearted, the Crusader King, gives us an excellent demonstration of a one-degree Saturn-Black Moon conjunction in Libra square Mars. The complex royal family dynamics included conflict between his parents, Henry II and Elinor of Aquitaine, and between Richard and his father and brothers. His life epitomized the historical alliances and tensions between England and France. He remains one of the best-recognized kings of England. A modern political example is that of Francois Mitterrand, president of France for a record- breaking fourteen years. He has a close conjunction in Cancer.

With his conjunction very close in Scorpio, trine Neptune in Pisces, Emanuel Swedenborg was a successful Swedish scientist and inventor who experienced a mystical awakening. He published numerous works in experiential theology explaining esoteric meanings of Christianity. The New Church and the Church of the New Jerusalem were founded on Swedenborgian ideas, which also influenced many thinkers such as Blake, Yeats (also with Black Moon and Saturn in Libra), Emerson, Baudelaire, Balzac, Jung and Helen Keller. John Chapman, better known as Johnny Appleseed, handed out Swedenborgian literature and apple seeds across the U.S. Henri Corbin wrote a study of Swedenborg's affinity with esoteric Islam. With activated psychic abilities, Swedenborg also experienced and wrote about visitations from planetary beings such as Mercury and Jupiter. This seer was most interested in the interface of spirit with matter.

The Outer Planets: Uranus, Neptune and Pluto

As we head out into the territory of the outer planets, we find that Black Moon conjunctions are experienced by a select group within a generational context. Uranus, Neptune and Pluto define generational tasks indicating changes in human consciousness as outmoded ideas to be radically altered (Uranus), dysfunctional areas needing clarity and refinement (Neptune) and repressed issues needing to be brought to the surface and transformed (Pluto). A Black Moon conjunction with an outer planet can imply an extreme experience of a generational theme that has the potential of being meaningful and making a mark at the high or low end of the spectrum of possibilities. Otherwise, an individual's experience can be internal and specific in a way not obvious to others. Either way, this does not promise an easy personal life. There is a double dose of the impersonally personal experience, often lifted to transpersonal dimensions.

Using the mean Black Moon we can zero in on its transits. True Black Moon moves in and out of conjunction repeatedly before and after the mean Black Moon. Please allow for this as we track the conjunctions in the following survey of mean Black Moon with Uranus, Neptune and Pluto. This line of research leads into fascinating discoveries, the scope of which cannot be fully plumbed here. We will focus this overview on the twentieth into twenty-first centuries. Refer to the Black Moon through the zodiac interpretations as needed.

Uranus

Black Moon Lilith joins Uranus for a few months in each successive sign placement, placing her special mark on certain representatives of that generation. Note the decade by decade rhythm of

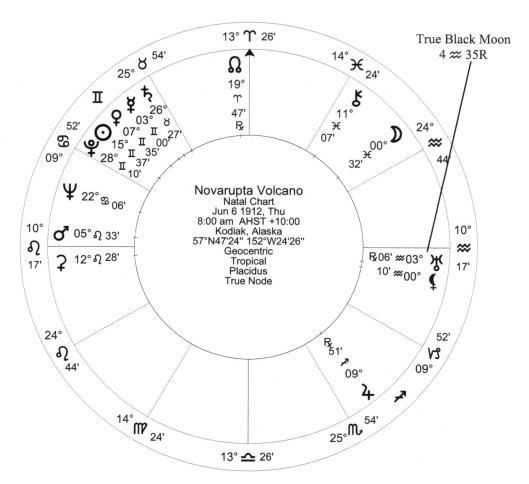

True Black Moon
4 ≈ 35R

Novarupta Volcano
Natal Chart
Jun 6 1912, Thu
8:00 am AHST +10:00
Kodiak, Alaska
57°N47'24" 152°W24'26"
Geocentric
Tropical
Placidus
True Node

the conjunction, every ten years. Uranus the Awakener has an affinity with the Black Moon as Dakini. Dakini is the cosmic feminine in Tibetan Buddhism, a multi-dimensional phenomenal energy carried by all women and many men. It is also experienced at many levels in one's inner spiritual journey. Dakini does whatever it takes to "blow your mind" beyond rational concepts so that pure consciousness awakens.

Uranus is the planet of the higher mind. Some of these people are outstandingly brilliant since both Black Moon and Uranus can endow a fine mind with rare insights. The gifts come with a price, potentially hard on the body. With this electrically-charged conjunction, the Uranian revolution of one's generation is uniquely inspired but potentially unsettling or misinterpreted, easier in some signs than others. The impact of this conjunction may be more fully channeled after the Uranus mid-life opposition.

1912—Uranus-Black Moon in Aquarius

With Uranus in its ruling sign, the conjunction here epitomizes its stormy potentials. Although true Black Moon was already in this sign, it was June 6, 1912 that mean Black Moon entered Aquarius to join Uranus and oppose Mars. Lilith demonstrated her chthonic power. It was on that very day that the massive, violent and devastating eruption of Novarupta Volcano in Alaska began. Over the next three days it dumped tons of toxic sulfuric ash thousands of miles away, its huge dark cloud eventually traveling across the globe to Algeria. A knee-high layer of ash buried the nearby town of Kodiac and the resulting environmental catastrophe killed wildlife over large areas, caused disease and economic disaster, and instigated the creation of Katmai National Park. Uranus retrograded into Capricorn in early September, effectively ending the conjunction.

Folk singer Woody Guthrie, whose songs addressed pressing social issues, is an excellent example of the best kind of change agent to be found with the conjunction very close in Aquarius. Master television chef Julia Child was also born at this time. Austrian Heinrich Harrer was born in July 1912. This champion skier and mountain climber, who joined the Nazi party in the late 1930s, was captured by the British in India while on a mountain climbing expedition. He escaped and famously spent *Seven Years in Tibet*, becoming friend and tutor to the young Dalai Lama.

It was the true Black Moon-Uranus conjunction that counted in the life of Rosa Parks, the gutsy black woman who famously refused to give up her seat on the bus to a white person, a pivotal moment for African American rights in the U.S. Spread across the sign of Aquarius in her chart we find Uranus, Mercury, Sun and Black Moon square Saturn in Taurus.

1922—Uranus-Black Moon in Pisces

Pisces is a sign of high ideals and sorrowful awareness of the suffering on this plane of existence. The martyr, the savior, the escapist or the medium can be an exaggerated outcome of this placement; true compassion is the blessing. Addictive problems can be a common strategy to dull the extra-sensitivity. This was an extended conjunction as the two flowed in tandem for the whole first half of the year, from January through Summer Solstice. The Sun's presence created a punctuation point as February turned into March. The conjunction peaked in late April into May. These natives experienced, and often led, the trends and dreams of their time.

Mentioned earlier in the Black Moon listings were beat writer Jack Kerouac and Israeli leader and Nobel Peace Prize winner Yitzhak Rabin. We find Italian designer Pierre Cardin in this group, and Judy Garland had the conjunction on her Midheaven.

1931-1932—Uranus-Black Moon in Aries

From November 1931 through March 1932, with waxing and waning effects before and after, Uranus-Black Moon in Aries impulsed its entrepreneurial spirit in the midst of the U.S.

depression. Pluto was square Uranus, a highly significant turning point on the global level and in collective consciousness. How does this relate to the return conjunction in 2011 of Uranus-Black Moon, during which Pluto is again square, this time in Capricorn?

A long list of famous people who made a large mark on the world stage were born during this Uranus-Black Moon square Pluto. Only a few are mentioned here. In February 1932, iconic actress Elizabeth Taylor and equally iconic country singer Johnny Cash were born a day apart, with an exact Uranus-Venus conjunction with Lilith. Edward "Ted" Kennedy was born earlier that same week, an enduring politician with a powerful scandal and tragedy-ridden Lilithian family. Two key women in his life, mother Rose and wife Joan, both had Sun-Black Moon conjunctions (see Leo and Virgo sections in the Black Moon listings).

Italian writer Umberto Eco was born in January with the conjunction on his Midheaven. American novelist John Updike, born in March, had Mercury in his conjunction. Another well-known country singer, Loretta Lynn, was born in April with Black Moon conjunction Sun and Mercury, as well as Uranus in Aries (see chapter three). Quirky original singer Tiny Tim of "Tiptoe Through the Tulips" fame was born two days earlier.

Puerto Rican actress Rita Moreno is noteworthy. Born in December 1931, the Uranus-Black Moon conjunction was in two strong planetary patterns. It was the hub of a T-square with Pluto in Cancer opposite Saturn-Moon-Venus in Capricorn, as well as one point of a grand trine in fire signs with Sun in Sagittarius and Jupiter in Leo. Winning an Academy Award at age thirty for best supporting actress in *West Side Story*, she went on to win Emmy, Grammy and Tony Awards, the first and only Hispanic actress to win all four awards.

1941—Uranus-Black Moon in Taurus, but not in Gemini

In October 1941, Uranus retrograded from Gemini into Taurus to continue its potent conjunction with Saturn, now joined by Black Moon Lilith. Chaos ensued with earth-shaking repercussions that required profound reevaluation. The old order was swept away. The global financial system was redesigned (the one that broke down in 2008!). This potent trio met in late Taurus, hunkering together like the three witches in *Macbeth*, with Saturn and the Black Moon conjunct at twenty-six degrees in alignment with Algol, the Lilith star, on Halloween. This triple conjunction persisted through November.

Home decorator business magnate Martha Stewart was born in this time. Right in the thick of the October-November alignment we find the singing duo, Paul Simon and Art Garfunkel (born just weeks apart), and Jesse Colin Young, lead vocalist with the Youngbloods and composer of the 1960s classics "Get Together" and "Darkness, Darkness" (a Lilith title). Preacher/politician Jesse Jackson was born with the triple conjunction emphasized by an exact conjunction of Moon and Black Moon.

Black Moon moved into Gemini the first week in December, just in time for the Pearl Harbor incident that propelled the U.S. into World War II. When the Saturn-Uranus conjunction moved across the cusp, from Taurus into Gemini, the Black Moon was too far along for either of them to catch up. So there was no Uranus-Black Moon conjunction in Gemini during that cycle.

1951—Uranus-Black Moon in Cancer

The conjunction of this generation was at its peak in October 1951, just as Uranus was stationing retrograde at 14 Cancer, the zodiac point that aligns with Sirius, the star of Isis. The conjunction stayed in operative range through Winter Solstice. Recall the discussion of Black Moon in the Moon's sign of Cancer, challenging to this typically security-seeking sign. Neither Uranus nor Black Moon offers a standard of security, least of all on the outer level. This alignment is a clannish one, with deep memory of Earth's pre-history, Lemurian mer-folk and the motherships from other star systems. Many of these souls know just how Sirius they are. They are great weavers of community through special personal bonds with "their" people, their tribes, their cosmic kin.

Famous musician Sting, former member of The Police, was born the same week in October as fellow musician Bob Geldof from The Boomtown Rats, who organized Live Aid and Live 8. In the July 1951 group we find genius comedian and actor Robin Williams and Oscar winner Anjelica Huston, popularly famous for her portrayal of Morticia Addams—a Lilith kind of character! Singer Chrissie Hynde of The Pretenders and Mark Hamill, a.k.a. Luke Skywalker, were born in the fall.

Neptune in Libra was square Uranus during the conjunction of 1951, so this is part of a larger cycle. Interestingly, when Uranus-Black Moon was in Cancer during 1872, Neptune was square in Aries. Philosopher, mathematician and social critic Bertrand Russell was born in that round.

1961—Uranus-Black Moon in Leo and Virgo

The conjunction in Leo was toward the end of the sign, strongest in August through October. A solar eclipse on August 11 opened the door to a peak week of influence that lasted through that whole Moon cycle, with the North Node tweaking Uranus-Black Moon. This can be an entirely outrageous combination: highly individual creativity that thrives on provocative expression and response. Depending on how open the heart is, we find geniuses and charismatic leaders as well as eccentrics and thrill-seekers. Many express this conjunction in a quieter way, on smaller stages than the global one.

The Leo conjunction is seen in the lives of Barack Obama and Lady Diana Spencer, two dramatic examples on the political stage. In the performance world Melissa Etheridge and Michael J. Fox both have the conjunction, with Mars added in. On the last round in 1872, artist Aubrey Beardsley had a Black Moon-Uranus-Mars conjunction.

Some souls took advantage of a very small window of opportunity when Black Moon joined Uranus in the specialty sign of Virgo, emphasizing the precision of that earth sign. Just around Halloween and All Saints Day, both Uranus and Black Moon stepped, Sphinx-like, across the royal star Regulus and across the Lion-Priestess cusp. Uranus stationed at 0 Virgo and the conjunction held until it retrograded into Leo in early January 1962. There is a deep ecological response to Earth energy in this Virgo position, and there are some rare specialists in these fields of endeavor. This group may have come in for the Harmonic Convergence in 1989, during which there was a multiple conjunction right at the sphinx-like juncture of Leo-Virgo.

The short-lived Virgo conjunction is shared by actress Meg Ryan, *Karate Kid* actor Ralph Macchio and Indian writer Arundhati Roy, the 1997 Booker Prize winner for her novel *The God of Small Things*—a good Virgo title. Sultry songstress k.d. lang, has the conjunction prominent on her Moon and Ascendant. Nadia Comaneci, the Rumanian gymnast who won five gold medals and was the first ever to get a perfect 10 score in Olympic gymnastics (seven in all), shows the exquisite technique of Virgo. She re-popularized gymnastics, inspiring a whole new generation of talent in the sport.

1971—Uranus-Black Moon in Libra

Mean Black Moon joined Uranus in Libra in June 1971, gearing up for their closest meeting from August through October, and waning as the year ended. Chiron plugged in from its Aries opposition, adding an extra charge during those peak months, highlighting the dialogue of I and Thou in person and in peace. This generation is inventing new ways regarding relationships.

This generational conjunction produced high-charged Puerto Rican singer Ricky Martin, rapper Snoop Dog, actress Winona Ryder and blue-grass country singer Alison Krauss. Two record-breaking athletes are in this elite group. Tennis super star Pete Sampras and cyclist Lance Armstrong, who won the Tour de France seven times in a row, including after cancer surgery.

1981-1982—Uranus-Black Moon in Scorpio and Sagittarius

The Scorpio generation can be especially interested in the occult, with a great capacity for regeneration and self-transformation. Roger Federer, discussed in the Moon section, is just one example of those born with this conjunction in summer-fall 1981. Israeli-American actress Natalie Portman is another. Tennis star Serena Williams was born two days after mean Black Moon moved into Sagittarius, but her true Black Moon is conjunct Uranus in Scorpio.

Uranus moved into Sagittarius in mid November, a month or so after the Black Moon entered the same sign. The conjunction continued to be close just a bit into January 1982, and then the Black Moon started picking up on the Neptune vibe (see Neptune section). We find several in this group helping to set new bars of accomplishment and extend the international sense of citizenship. Argentinean tennis star David Nalbandian is a sports representative. Sagittarius, as well as Black Moon, can tend to exaggerate and overextend, leading to downfall, so if Black Moon in Sagittarius doesn't "walk the talk," it gets into trouble sooner or later. Uranus makes it obvious. This Sagittarius group includes troubled top teen vocalist Britney Spears, who shows many facets of Lilith with both Mercury and her Sun joining the conjunction.

1991—Uranus-Neptune-Black Moon in Capricorn

The Uranus-Neptune conjunction was a cosmic moment not to be missed. True Black Moon made forays into Capricorn as early as January, and mean Black Moon stepped in at the end of April, moving into alignment with the waxing conjunction of Uranus and Neptune in August, September and October. The presence of the North Node intensified the karmic implications. Persons in this elite group often seem mature beyond their years. This is a multi-nuanced alignment profoundly engaged with the Cosmic Crone and Earth nature hierarchies in process of re-matrixing the grids and sacred geometries. Further implications of this alignment will be revealed as Pluto transits mid Capricorn degrees in 2013-14.

2001—Uranus-Black Moon in Aquarius

Black Moon came into Aquarius in late November 2000, first visiting Neptune and then moving on to join Uranus more closely in June, particularly after Summer Solstice and through August, immediately preceding the pivotal date of 9/11. The toxic ash from that devastation recalls the previous 1912 conjunction accompanied by the eruption of Novarupta.

2011—Uranus-Black Moon in Pisces and Aries

This conjunction immediately precedes another pivotal moment: the 2012 shift at the Capricorn solstice, long noted in the calendar of the Mayans as a turning point in human and planetary evolution. Black Moon connects with Uranus as it is transitioning from Pisces to Aries. The conjunction becomes strong as the year opens, peaking in February and continuing into March, when both factors cross over into the fresh territory of Aries along with the Sun. They march along together until Summer Solstice, then Uranus slows down to retrograde, while Black Moon moves further into the (black) virgin territory of Aries. Note that three of the seasonal turnings of the annual cycle have been mentioned as active in this scenario, as the cosmic wheel turns in the 26,000-year precessional cycle.

Neptune

Fascination and seduction, veils of illusion, confusion and delusion, as well as the mystical, magical and spiritual are keywords shared by both Neptune and Lilith. Put the two together and we have extraordinary, practically extraterrestrial results that can be at one extreme or another. Neptune refers to the ideals of a generation, its aspirations prone to go off track into less than ideal byways or situations that require sacrifices. Black Moon can darkly thicken the veils of illusion or part them, revealing the worst or best of the generational aspirations. If the selfhood isn't strong enough, one cannot easily contain the energy of this conjunction. It may be hard for the person, not to mention to the public, to see clearly what elusive star is being followed. The person can project an aura that is beyond his or her personal self, that symbolizes something meaningful for the times. The rhythm of this conjunction creates an alternating pattern through the course of Neptune's 165-year orbit. Neptune spends fourteen years in a sign. There will be two conjunctions at the beginning and end of one sign, followed by a single conjunction in the mid-degrees of the next sign.

1907—Neptune-Black Moon in Cancer

Remember the group of iconic movie stars mentioned in the Jupiter section that were born during the 1907 conjunction? Katherine Hepburn, Sir Lawrence Olivier and John Wayne each brought their own special aura to the big screen, with Neptune's filmy glamour amplified by Jupiter, all opposition Uranus-Mars, marking their award-winning popularity. These three were all born in May, at the beginning of the conjunction that lasted into October. Others born at this time were Mexican artist Frida Kahlo, British writer Daphne DuMaurier of *Rebecca* fame, and *Far Memory* reincarnation novelist Joan Grant.

1916-17 and 1926—Neptune-Black Moon in Leo

The Leo conjunction came into play in November and December 1916, waning out of effective range by early March 1917. Italian auto maker Ferruccio Lamborghini , Kirk Douglas and Marilyn Monroe were born in this section of time. MM had this charismatic conjunction in her first house. Born in January 1917, Maharishi Mahesh Yogi, founder of Transcendental Meditation, illustrated the spiritually-oriented side of Neptune and Black Moon. Were any scandals connected to this guru?

In 1926 the Black Moon caught up with Neptune in March through May.

Science Fiction Hall of Fame writer Anne McCaffrey was born April 1 with an exact mean Black Moon-Neptune in Leo on her Ascendant. The first woman to win a Hugo Award for science fiction (Best Novella in 1968), she created multiple fantasy worlds, including the popular *Dragonriders of Pern* and *Crystal Singer* series. Music is a big theme in her extensive bibliography, as are esoteric themes such as telepathy and other experiences of altered consciousness. Hers was an active Neptune-Black Moon, opposition Jupiter-Venus in Aquarius and square Moon-Saturn in Scorpio—a very intense fixed T-square. She was married for twenty years before divorcing and moving to Ireland as a single mother of three children. She lives in her self-designed home called Dragonhold-Underhill. Her fantasy novels are Lilith lexicons.

Also born in this time were television actor Andy Griffith, Playboy bunny lover Hugh Hefner, jazz legend Miles Davis, and Michio Kushi, founder of Erewhon Natural Foods, the *East-West Journal* and the Kushi Foundation.

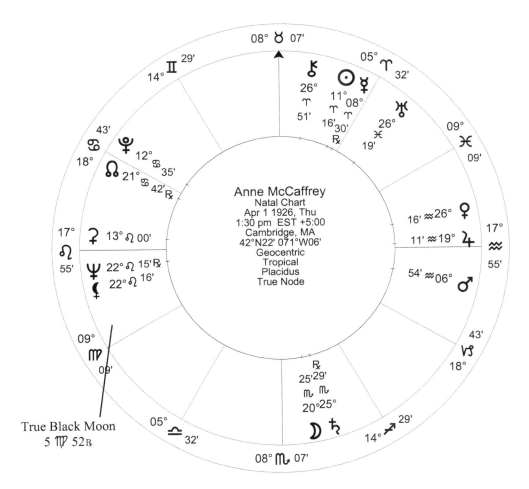

True Black Moon
5 ♍ 52℞

1935—Neptune-Black Moon in Virgo

The Black Moon went into Virgo in April 1935, reaching a five-degree range with Neptune during the months of June through October. As previously noted, the 14th Dalai Lama, born in 1935, has Black Moon closely conjunct Neptune and the Moon, as befits his spiritual status. His astro-twin, Candy Barr, lived long in the Neptunian drug and glamour world of the exotic dancer, and wrote poetry in prison. Woody Allen was born with a wide conjunction of Black Moon and Neptune. Opera stars Luciano Pavarotti and Donald Sutherland both had Venus opposite this conjunction. Singers Julie Andrews, Johnny Mathis and "Great Balls of Fire!" Jerry Lee Lewis were all born the same week in May.

1944-45 and 1954—Neptune-Black Moon in Libra

Black Moon entered Libra in November 1944, joining Neptune in the early degrees through February 1945. Well-known persons of this time include Reggae king Bob Marley, actor Tom

"Magnum" Selleck and Mia Farrow, actress and UNICEF Goodwill Ambassador.

From Spring Equinox through May 1954, the conjunction was again in play. This younger group includes superstar John Travolta and activist/documentary filmmaker Michael Moore, who won an Academy Award plus many more for *Bowling for Columbine,* which explores America's predilection for gun violence.

1963—Neptune-Black Moon in Scorpio

Black Moon was in this dark and deep sign from late August into mid January 1964, with its peak in July through October. In addition to previously mentioned actors Johnny Depp and Brad Pitt, we also find Lisa Kudrow of *Friends* and actor Greg Kinnear with this Scorpio conjunction, along with singers Tori Amos and superstar Whitney Houston. With an exact and exacting conjunction, Houston has had her share of addictions and rehab. *Paper Moon* actress Tatum O'Neal has the conjunction on her Sun-Mercury. At that time at age ten, she was the youngest Oscar winner ever.

1972-73 and 1982—Neptune-Black Moon in Sagittarius

From mid November 1972 into April 1973, Black Moon and Neptune joined in this broad-ranging sign. Charismatic actor Jude Law has a close conjunction, possibly overshadowing his Ascendant. Illusionist David "Street Magician" Blaine was born during the end of this phase. His daring stunts grabbed public attention and some skepticism, as he was frozen in time, buried and drowned alive, fasted for forty-four days, stood on a 105-foot pillar for thirty-five hours, and other such tricky things.

In 1982 the Black Moon came into Neptune's range by late March, reaching peak conjunction in May and until Summer Solstice. Prince William, with his seemingly endless layers of Black Moon, was born on this Solstice and has the ineffable Neptune-Black Moon on his Ascendant. Actress Kirsten Dunst and *American Idol* winner Kelly Clarkson are on this list.

With Sagittarius ruling sports, we can expect to find players in this list. The tennis world gives us several: Belgian Justine Henin and Spaniards David Ferrer and Tommy Robredo. Italian tennis player Flavia Pennetta is an example of someone born in the in-between months, as Black Moon moved from Uranus to Neptune in February-March, with Black Moon at the midpoint of the two planets.

1991—Neptune-Black Moon in Capricorn

The peak of this alignment came in late July through October, as Black Moon emphasized the waxing conjunction of Uranus-Neptune. Princess Mako of Japan was born in late October, with the North Node present in the conjunction, a call to destiny from the dark goddess.

2000-2001 and 2010—Neptune-Black Moon in Aquarius
Starting in November, the strength of the year 2000 conjunction lasted to Spring Equinox 2001, then held at the Neptune-Uranus midpoint before going on to fully engage with Uranus. In 2010 the conjunction was strongest in May and June.

2019—Neptune-Black Moon in Pisces
Let's mark our calendars for late summer into fall of this year for this conjunction in Neptune's own sign.

Pluto

Whatever scientists call it—planet, dwarf planet or plutoid—this remote dancer in the dark will do what Pluto does. Astrologers know not to ignore it. A connection between these two dark energies, Pluto and Black Moon, taps into something much larger than the personal and is a conjunction shared by a group of people. It may need a personal planet to pull it into individual significance.

Pluto's highly elliptical 248-year orbit is irregular; it moves through some signs at a leisurely 30-year pace, then quickens as it moves closer to the Sun. At its perihelion, while in Scorpio, Pluto crosses Neptune's orbit and is closer to the Sun for twenty years. The most recent period was 1979-1999. From several conjunctions with the Black Moon in a sign to just one, we see the fast pace of Plutonian evolution in this era. There are more Pluto generations by sign alive together now, in an era of high population density, than at any other time in the last couple hundred years at least.

1906-1907—Pluto-Black Moon in Gemini
This conjunction occurred several times in the late nineteenth century as well, impacting the Neptune-Pluto conjunction that inaugurated the twentieth century and a new 500-year cycle. The triple conjunction of 1898 saw the births of Israeli leader Golda Meir, dramatist Berthold Brecht and stigmatic Romanian mystic Therese Neumann.

The last Gemini conjunction was in close orb from December 1906 into early April 1907. Mircea Eliade is a good example of the paradigm-altering research and perspectives contributed by many born at this time. Religious historian, writer, philosopher and professor, his interpretation of the "eternal return," and the relationship between sacred and profane, opened a new reading of religious experience through the ages. W.H. Auden, recipient of the 1949 Pulitzer Prize for "The Age of Anxiety," left a significant poetic legacy. Soichiro Honda left a name brand that continues to drive us around in true Gemini fashion. Leonid Brejnev was the longest-lasting political leader of the USSR in recent history, serving as general secretary of the Communist Party for eighteen years. Sir Laurens VanDerPost was mentioned in the Gemini section for his multi-faceted contributions in literature and politics. Albert Einstein was publishing his seminal work in this window of opportunity.

1916, 1925 and 1934—Pluto-Black Moon in Cancer

Black Moon met Pluto for three months beginning in mid February 1916 and went on to conjunct Neptune. In 1925, the conjunction returned into orb from February through May, and in 1934 from late March through June. As earlier discussed, Black Moon is not easy in Cancer, and Pluto doesn't make it any easier. Issues in relation to home, home land and family can be life-defining beyond the ordinary, archetypal in their expression. Multiple generations of this placement give us interesting examples.

From 1916: Oscar winners Gregory Peck and Olivia DeHaviland, as well as international violinist and conductor Yehudi Menuhin.

From 1925: Former First Lady Barbara Bush, wife of George Bush I, has an exact Mars-Black Moon conjunction opposite Moon-Jupiter in Capricorn. Her family was filled with men of power. In addition to great performances, famed actor Paul Newman also gave us Newman's Own food line, a Cancer thing to do. From bad boy to converted Black Muslim minister, Malcolm X stood for empowerment of his people. Astrologer Linda Goodman introduced many to the stellar art, with her *Sun Signs* becoming the first New York Times best-selling astrology book. I'm not the only one who was turned on to astrology by this author.

From 1934: Actor Richard Chamberlain, from television's *Doctor Kildare* went on to miniseries (*The Thornbirds)* and film (*The Last Wave)* as well as the theater. Oscar-winning director Sydney Pollack left a large list of memorable films. Famed actress Shirley MacLaine "came out" as a believer in reincarnation. Designer Giorgio Armani remains a leader in classy men's fashion.

The Canadian Dionne quintuplets (Annette, Cecile, Emilie, Marie and Yvonne), a rare set of five identical Gemini girls, born two months premature in 1934, are the first known to survive infancy. Born to a poor couple in the Canadian back woods who already had four children, the unusual birth was not considered a blessing but rather a burden. The girls were used and abused by both family and government for fame and gain. The Chicago World's Fair administration wanted to put the babies on show. The government stepped in, making them wards of the king and providing a trust fund for the girls under the Dionne Quintuplets' Guardianship Act. The huge publicity meant they were reared basically as a tourist attraction, first by a clinic, then by their parents. A very close Pluto-mean Black Moon conjunction in late Cancer was square Venus-Uranus in Aries. Their Suns were conjunct Chiron as well, adding another note to their unusual, even unprecedented experience, variously described as exploitation, melodrama or tragedy—all Pluto words. They all left home at age eighteen to pursue their own lives. As to their fates: Emilie died during an epileptic seizure at age twenty. Marie died of a blood clot at age

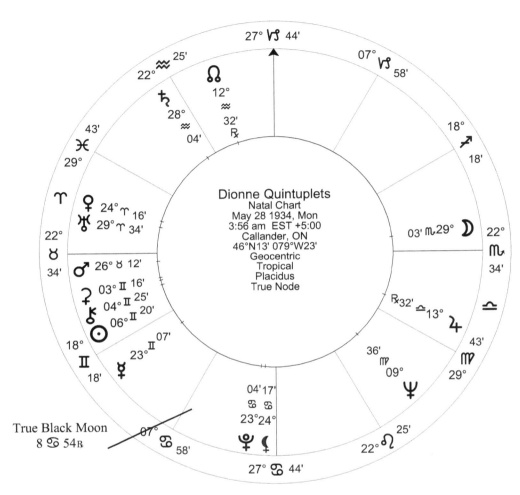

Dionne Quintuplets
Natal Chart
May 28 1934, Mon
3:56 am EST +5:00
Callander, ON
46°N13' 079°W23'
Geocentric
Tropical
Placidus
True Node

True Black Moon
8 ♋ 54ʀ

thirty-five after a troubled adult life. Yvonne died of cancer at age sixty-seven. As of this writ-ing, Annette and Cecile, both divorced with children (Cecile had twins), live quietly near Mon-treal, estranged from their parents, finding their own lives out of a story for the ages.

1943 and 1952—Pluto-Black Moon in Leo

The first conjunction occurred in late May through September 1943. The Black Moon returned to Leo at the end of March 1952, becoming strong in late August through November. Big heart, heroism, humility and honor are stellar qualities expressed by this very creative and re-creative generation. The ego learns to take a back seat to the soul, not always graciously, but often in the loveliest ways.

In the Leo section, we saw a list of 1943 natives who have the rare Jupiter-Black Moon-Pluto conjunction, some with Mercury added: Mick Jagger (on Sun), Robert DeNiro and killer's accomplice Caryl Ann Fugate.

The 1952 group includes Christopher Reeve (already mentioned with Black Moon on a Leo Ascendant); actor/dancer Patrick Swayze, who wrote and sang "She's Like the Wind," from his break-out movie, *Dirty Dancing*; and charismatic actress Isabella Rossellini. Medical intuitive and *Sacred Contracts* author Caroline Myss has this conjunction, as does inspirational lecturer Marianne Williamson, a self-described "provocateur" who founded The Peace Alliance calling for a U.S. Department of Peace. Joni Mitchell has the conjunction wide, square her Sun-Mercury in Scorpio. Record-breaking American tennis star Jimmy Connors was born this year with the Sun in Virgo.

1961-1962 and 1971—Pluto-Black Moon in Virgo
In October 1961, the Virgo conjunction came into play, hitting its stride from mid December into March 1962. It is as if these people seek constant communion with Earth energy. They can have super-skills and specialty areas, healing abilities and a desire to be of service. Those rare multi-Aquarians born in January-February 1962 who have up to five or six planets in this sign have this conjunction as well. Singer Sheryl Crow is in this Aquarian group, as is Garth Brooks, the highly successful country singer who crosses into pop music. Beyond the Aquarians, we find rocker/actor/activist Jon Bon Jovi, plastic-face actor Jim Carey and animal lover Steve "Crocodile Hunter" Irwin, famously killed by a stingray, an especially interesting soul contract.

The second visit was in April and May 1971. Members of this special 1970s group are colorful neo soul singer Erykah Badu (wide conjunction opposite her Moon), Japanese teen idol Seiko Matsuda (with the conjunction on her Ascendant as discussed in the Virgo section) and Sofia Coppola, daughter of Francis Ford Coppola, who was nominated for two Academy Awards for screenplay and direction of *Lost in Translation* in 2003. She won for the screenplay.

1980—Pluto-Black Moon in Libra
This conjunction took place in mid-Libra, most closely from August to December 1980. Relationship is a hallmark of their lives, most profoundly in the sense of a right relationship with Earth, with others, and with ourselves in work, play, love and life.

Born a day apart in December, singer Christina Aguilera and actor Jake Gyllenhaal both made an early impact on the world stage, along with actor Ryan Gosling. Top tennis stars Martina Hingis of Slovakia and Venus Williams of the U.S. fit into this group. Venus Williams has a ten-degree span between Pluto and Black Moon, with her Ascendant in between. You can peep at Kelle Marie, Welsh erotic actress and porn star, on various websites, especially her own. She has a very close conjunction!

1990—Pluto-Black Moon in Scorpio

Pluto was relatively racing along through the signs at this point, having hit perihelion in 1989. Effective from late February to early May 1990, this Scorpio conjunction is very intense as Pluto and Black Moon, both deep and dark, were in the deepest and darkest sign. This was also during the time of the Uranus-Neptune-Saturn conjunction in Capricorn. There were some very special people born in this group, which we could call the Harry Potter generation, very aware of the dark side, of life and death, capable of great regenerative capacity. Emma Watson, who plays Hermione Granger in the Harry Potter movies, was born with all of the above. Her real life has mythic resonance. She attended the prestigious Dragon School in Oxford and auditioned for the Harry Potter movie role at age nine.

1999 and 2008—Pluto-Black Moon in Sagittarius

On the threshold of the new millennium, Black Moon Lilith went into Sagittarius in June 1999, hitting the peak of its conjunction with Pluto in July through September. Many in this group are reborn visionaries, bards and master teachers. They travel far. Thus it was in effect during the powerful total eclipse on August 11, 1999, when the conjunction was pretty much exact at eight degrees. The 2008 conjunction was briefly in play in November at the far edge of Sagittarius, just before Pluto changed signs, moving into Capricorn.

2009 and 2018—Pluto-Black Moon in Capricorn

The Black Moon then caught up to Pluto in early January 2009, during Barack Obama's inauguration, the conjunction returned, lasting into March. Another conjunction will occur in this sign in April, May and June 2018. The power of Earth as a living, evolving entity are made evident as we embrace the sacred geometries that create the matrix of life.

The New Solar System

I am not at this time prepared to comment on the Black Moon in relation to the newly-discovered planets, dwarf planets and trans-Neptunians. However, we can intimate from mythic resonance that Black Moon as Lilith has a significant relationship with Eris, goddess of discord, and Sedna, who lives as deep in the Arctic Sea as in the outer depths of the solar system. As elusive as the TNOs, we can imagine that she is most willing to run wild with the Centaurs and keeps company with her many mythic brothers and sisters in the asteroid belt. In her four astronomical expressions—Black Moon, Dark Moon, asteroid and star—Lilith spans multi-dimensions, from inside Earth out 105 light years, the distance to Algol. Like a note at different octaves, one Lilith resonates with the others. Black Moon, then, vibrates to influences sounding through interstellar space.

Chiron

When infused with Chiron's quantum healing quest, subtle and perplexing experiences lead to deep inner transformation and unusual expertise conveyed energetically. Chiron has an irregular orbit, so this conjunction will have an irregular pattern. In its fifty-year orbit, Chiron swings between Saturn and Uranus. When it is closer it moves through the signs more quickly, spending two years in Virgo at perihelion. When out near Uranus, its transits are longer, seven years in Aries. It tends to have regular visits with the Black Moon in Pisces and Aries, but won't always conjunct the Black Moon in its shorter transits. Here is an outline of the exact conjunctions in the twentieth into twenty-first centuries. Give two months on either side for effective orbs (longer transits are indicated), and remember that the true Black Moon can be sporadically effective for a longer span.

May 1903, 21-22 Capricorn
July 1913, 15-16 Pisces
January 1923, 14 Aries
May 1944, 10 Virgo (strong into February 1945, when it was in early Libra)
October 1956, 7 Aquarius
May 1976, 23-29 Aries
May 1986, 20-21 Gemini (May-October)
1998 (wider than 10 degree orb in Scorpio, but close to conjunction)
June 1999, 1 Sagittarius
July 2010, 0-1 Pisces
February 2020, 3-4 Aries

Again we encounter the 1976ers, a special group with their Black Moon-Jupiter joined by Chiron. There is another remarkable and unique band of 1956ers with Black Moon-Chiron in Aquarius opposed to Uranus in Leo. This conjunction was fully in effect from October through December. These people are like lightning rods, plugged in to the outer bands of the atmosphere, downloading and upgrading with passing transits. They can have systems with alternative and refined wiring that can lead to both strange symptoms and strong healing energy.

The Lunar Nodes

Black Moon Lilith and the Lunar Nodes have an affinity, through astronomy as well as mythic imagery. The Nodes are called the Dragon Head and Tail; Lilith has an ancient serpent lineage. Both are invisible points with mean and true positions mathematically involved with the intricate dance of Earth-Moon-Sun. The Nodal cycle is twice as long as the Black Moon cycle through the zodiac.

The Nodes are karmic points that indicate the flow of fate or destiny and the growth edge of consciousness development, which we can cooperate with or resist. Like the Nodes, the Black Moon also indicates deep forces at work beyond our conscious understanding that illumine soulful awareness of underlying spiritual currents. The Nodes are a polarity of continual interplay—the one implies the other; however, there can be an emphasis determined by planetary conjunction. A South Node conjunction would refer to some current of energy carried from the past into current life, while the North Node conjunction suggests some aspiration for achievement or karmic stipulation in this lifetime.

A North Node conjunction suggests a karmic stretch, a "compulsion to experience life," writes Vedic astrologer Komilla Sutton. The North Node pulls us into incarnation, is ambitious. The Black Moon may not be ambitious, precisely, but does imply some soul desire for embodiment. One can wade through dark, murky waters before surfacing to embrace the true quality of experience that satisfies.

The South Node, more instinctual, emotional and intuitive, is a natural release point when potential resistance or pull from the past that needs to be overcome to free the energy into expression. In Vedic astrology it reveals fears of the ego, a rejection of worldliness and materiality, indicating an essentially spiritual point when the darkness that hides the light is dispelled. These themes fit well with the meaning of the Black Moon to remove veils of ego identity toward spiritual illumination.

Perhaps people with Black Moon Nodes are agents of the goddess, whether for demonstrating or awakening some theme relevant to the collective consciousness, recognized or not in their lifetimes, or perhaps for some privately recognized experience. The presence of a planet on the Nodes channels the energy more specifically. The meaning of a Nodal-Black Moon conjunction is inscrutable. It leads to a compelling sense of destiny at work in life. One can probe with questions and queries, but essentially such a person has a life-long date with the dark goddess.

Lech Walesa, Polish leader and devotee of the Black Madonna, has Pluto and Jupiter conjunct the Black Moon-North Node, as do many of that 1943 group mentioned earlier. Like many of this mini-generation, Walesa strove for a philosophically-based form of political leadership, willing to stir things up to create afresh, or for an expression of some elusive ideal.

The chart of the first Radical Gay Faerie gathering held in the desert west of Tucson, Arizona on August 31, 1979 has the two Black Moon positions spanning half of Virgo with a very close mean Black Moon-Sun-North Node-Venus conjunction and Saturn nearby. A Moon-Neptune conjunction in Sagittarius adds a big dose of indulgence, adventure and high spirits, in square to the more sober true Black Moon and Saturn. The Sabian symbol for the Sun-North Node conjunction at 9 Virgo in the chart reads: "A modern expressionist artist paints a strange canvas."

Dane Rudhyar amplified this as, "Original genius of every individual soul unconcerned with collective values. Absolute, tradition-less self-expression." The Sabian symbol for the 6 degree Virgo mean Black Moon adds a further tone: "Excited children ride on a blatant, gaudy merry-go-round," indicating "The culture of pleasure as a transmuting force. Unfearing plunge into life." This group is creatively and socially radical, wild and wacky, compelled to incarnate a powerful unfettered essence of the feminine consciousness arising in the *anima mundi*. Always held in Nature, these gatherings continue annually in several states and countries. A member of this tribe offered one of the first real-life mentions of Lilith in my experience. I believe there is a particular honoring of Lilith at their gatherings. These artists also dance, prance and are members of various international orders of the Sisters of Perpetual Indulgence, a leading-edge order of queer nuns who take vows "to promulgate universal joy, expiate stigmatic guilt and serve the community," as their website states (thesisters.org).

Steven Spielberg (see Sagittarius section) and Andy Warhol both have a Black Moon-South Node conjunctions in Sagittarius. Both contributed iconic imagery to the collective. NASA, the American National Aeronautic and Space Administration, has a Black Moon-South Node conjunction in Aries, preparing them, a la Star Trek, to "go where no man has gone before."

Chapter Six

Lilith's Dark Mirror

A female tiger. She is magnificent, powerful. We treat her with respect, awe. She can hurt, but we are allowed to stroke her. She is surrounded by a round enclosure, trapped. A number of male cats come in and rape her. She is covered in blood after the second attack. After, she is left encaged, her heart destroyed. Anyone who approaches her is met by a terrible, hateful, warning snarl. She is dangerous, ferocious, destructive, defensive. Why such a royal upbringing to be led to this fate?

Thus Lilith entered my dreams. Women experience the challenge of Lilith in acknowledging, claiming and expressing their unique feminine co-creative essence in a culture that has been repressing this power for thousands of years. In Jungian terms, she will evoke a woman's shadow aspects, mirroring unconscious motivations, hidden potentials and spiritual capacities not initially recognized or expressed consciously. The journey to develop our spirituality and trust our inner strength and knowing is not easy, as we uncover layers of feeling, like archaeologists digging for lost worlds. Feelings of vulnerability, rage, despair, fears of rejection and isolation are stirred up, as we sift through the dirt covering the soul and come home to ourselves. Committing to these deeper spiritual aspects of Self may at times seem like betrayals to others. If we do not do this, we are betraying ourselves. The continual letting-go process may feel endless. Lilith takes us beyond these personal reactions to a deeper core of truth.

For men, Lilith is an aspect of the anima, as Jung named the female aspect found in the unconscious level of the male psyche. A strong Black Moon in a man's chart indicates a strong presence of this "inner woman," a guide on the spiritual path. As Mediatrix of the numinous and liminal, everything reflected in this inner mirror to the soul becomes dangerous, taboo, magical, constantly shifting and immensely alluring. Encountered in relationships with women and in their own soul journeys, Lilith challenges men to experience the feminine without projections based on conditioning, inadequacies, fears, or urge for control. Lilith cannot be controlled. It dishonors the feminine to even try.

Spiritual influences from more expanded realms often come with a sexual charge. As we are in the world of duality, our consciousness translates this joining with higher states of consciousness through sexual images and feelings, although the original idea is beyond gender. This is expressed in mystical writings in many religious traditions.

Flora Mask

Lilith in Relationship

Relationships are most powerful mirrors, reflecting and revealing previously unknown depths of self and other, both loving and unloving. It is said that "Love brings up everything it is not." Relationships that contain Lilith energy are initiatory, soul-to-soul meetings that open us into a deeper center of our experience, dissolving self-definitions and undermining our emotional comfort level. Can we really trust this energy that is unraveling our edges, plunging us into an emotional chaos? Can you trust yourself?

Liilth is highly erotic. Eros is the primal power of love, the intense passion that creates the universe through the power of attraction and desire, transforming chaos into order. Eros is most commonly known as Cupid, the naughty cherubic son of Aphrodite whose arrows cause wounds of love. Such a Valentine image of sentimental romantic love hardly does justice to the creative essence inherent in the erotic function. The term erotic is understood mainly in connection with physical sexuality, its pleasure of intensification and release. In essence, Eros seeks not release, but rather an ever-deepening intensity, letting go into a pure experience of sat-chit-ananda = existence-consciousness-bliss. True Eros focuses the life-generating dance of the cosmic masculine and feminine and the juicy, creative and re-creative tango of lovers. Erotic desire is profoundly transformational.

What is love? And what do we mean by falling in love, that delicious and dangerous state we both deeply fear and deeply desire? Eros is the desire for union in delight and passion, evoking new dimensions which open the heart and soul of lovers. The passion of love brings new worlds into being and reveals us to ourselves and each other. Eros is desire and the "drive toward union with what we belong to," adds Rollo May in Love and Will. "Eros seeks union with the other person in delight and passion, and the procreating of new dimensions of experience which broaden and deepen the being of both persons." Love opens a vision of the Infinite. Eros taps into the power of awesome realms, throwing a hot light into the shadows. Love is a plunge into the great unknown, a churning sea of hot, erotic, sensual/emotional chaos. It takes great courage and endurance to create and re-create love.

One of the original chaos theorists, Ilya Prigogine, defined two levels of chaos. The first is a passive chaos of equilibrium where the elements are so intimately mixed that no organization exists and the system just breaks down. Translating this into relationship, this is when the comfort level gets too cozy and predictable. Eros flees, and so does Lilith. When emotional dependencies build up in a relationship, the sexual fire may wane and we wonder where the passion went. Lilith comes in to remind us what we *really* want, often showing up as "the other woman" or the "femme fatale." When the emotional issues are worked through, deeper levels of intimacy can support a richer passion than ever. A woman may find herself ins uch a role.

The second kind of chaos is a turbulent one, active, hot and energetic, in which strange things may occur and where systems don't just break down; new systems emerge. This is the spicy tango, the passionate dance that brings Lilith to life and life to Lilith. Take dance lessons from Lilith!

There is often a necessary process of separation in the erotic relationship. Black Moon relationships require profound letting go. In separation the erotic tension guilds to a burning point. Sometimes—too often!—our passions contain neurotic aspects, manic and demonic attachments. We go slightly crazy in love. Some people go really crazy and commit crimes of passion. Through separation, even betrayal, we become aware that "love is blind." For the sake of our well-being and sanity, we must let go of all erotic compulsions: the self-centered sadness, pain, unrequited yearing and depressed longing that hold us back from the creative freedom of true love. "Always the ardour of love transmutes fear and compulsion into a higher freer type of love," wrote Carl Jung.

This erotic perception of Truth, through the agony and ecstasy of intimate relationship, leads ultimately to merging with the divine. Love is the source of consciousness, of personal and planetary change. Black Moon Lilith taps into the erotic dimension of our souls and we are transformed through the penetrating touch of that daimonic force.

When Black Moon is active in chart synastry, it may be the most compelling factor in the relationship. A Lilith relationship is a spicy tango or a psychological duel, which may spur spiritual growth, but sometimes in a rather cruel way. As I wrote in *Living Lilith*:

Such relationships are not about having a family, a simple love affair, or going into business together,although these things may happen. Lilith does not care about the Moon concerns for comfort level or meeting subjective personality needs. Nor does she bow to Venusian emotional pleasure as the prime mover of the relationship, though she doesn't dismiss such sharing. The main agenda behind a Lilith connection is soul growth and spiritual liberation.

Such relationships are ego-busters, requiring a willingness to connect from one's truest essence. Along the way Lilith is likely to free up sexual inhibitions so we can feel that primal passion of the "big bang," or the "big bloom," as philosopher Ken Wilbur calls it.

Black Moon Lilith challenges women and men to connect with their instinctive passion for life, for this natural force turns destructive when denied, unfulfilled, caged or exiled. Whether woman's shadow side or man's dark inner woman, when the deep waters of the unconscious are stirred, this is not sweet-tasting tropical punch. This is the bubbling cauldron of the three witches, the triple Lilith, made of bitter, psyche-purifying herbs and roots, flavored with owl droppings, dragon snot and serpent venom. It is the grit and gruel of inscrutable emotions brewing beneath the surface that are going to spice up the soul and burn a hole through the veil of our comfortable illusions about ourselves. Lilith awakens us to our unknown inner selves and urges us through the dark passageways where she lurks, disguises, tests. Like the Sphinx, she utters riddles in the circumstantial details of our lives. She hides the answers behind her Mona Lisa smile. She insists that we feel through, let go and surrender to the something deeper in us that is primary, the bedrock beneath the shifting sands. We look in her mirror and see our true selves. The ego quakes; the spirit exults.

Lilith leads to the reality of the psyche mediated by intuitive instinct, through which we have the capacity to expand more deeply into our wholeness, to resonate with loved ones and our community circles, and to spiral ever out into larger cosmic cycles. We dance with her in the web of dark matter, of the Dark Mother/Creatrix.

Lilith Speaks: A Dramatic Monologue

My first Lilith "research" came through dreams and creative expression. While creating a series of mythic dramas with Dragon Dance Theater, I played Lilith, developing a character to give voice to this dark goddess, to the pain of the female vital life force betrayed, suppressed, now to be acknowledged and redeemed. I blended material from my dreams, from the Biblical "Song of Songs," the Gnostic Gospels and other sources. From this work I learned that Lilith is the Tree of Life, offering the fruits, not simply of knowledge, but of true wisdom She offers direct experience of spiritual spaciousness and if we can flow with her beyond our fears from the past, beyond our fear of the unknown, beyond even our fear of death.

Focusing on this material for eight months was a profound experience that underscored other aspects of my life and relationships. Lilith lured me into deeper aspects of my unknown self. Creative work provided a channel through which to process an inner and outer transformation toward recognizing some darker emotions, acting from my personal center, expressing more fully my sexuality, and clarifying appropriate levels of intimacy in relationships. I began to give voice to Lilith in the form of a dramatic monologue created from this mythic theater work. The intriguing circumstances that occur around presenting this piece continue to educate me in her mysteries.

Today Lilith, the Cosmic Feminine, calls us all into her healing powers re-emerging from the depths of Earth itself for our shared awakening.

> *Lilith enters:*
> *You wonder: Who am I coming at you like your shadow?*
> *Like a bird in flight*
> *in the dark of the night*
> *"I am the first and the last.*
> *I am the honoured one and the scorned one.*
> *I am the whore, and the holy one.*
> *I am the substance and the one who has no substance." (2)*
> *"I sleep on the earth and I dance in the trees.*
> *I lie on the sands and I fly on the breeze.*
> *I walk in the sun and I drink with the bees.*
> *I sing with the rocks and I do as I please." (4)*
> *You have summoned me with your desires*
> *There are things you need to know about Love and Creation.*
> *What can I know about Love, you ask?*
> *I seduce, I destroy, I have no heart?*
> *Those are lies, lies they tell about me now,*
> *Lies they might tell about you later.*
> *I live in the Tree of Life,*
> *with the serpent in the roots and the Thunderbird in its branches.*
> *Why are you afraid of me, afraid of the tree?*
> *"What you see outside is what is inside of you.*
> *And what is inside of you is what is outside of you.*
> *It is visible and it is your garment.*

If you bring forth what is within you,
what you bring forth will save you.
If you do not bring forth what is within you,
what you do not bring forth will destroy you." (2)

To Woman:
"How beautiful are thy feet, O Daughters of the Moon!
In the joining of thy thighs there is a jewel." (1)
I am wild like the wind. Yes!
"Thy navel is like a round goblet filled with wine.
Thy belly is like a heap of wheat set about with lilies." (1)
I live in trees, I kiss serpents. Yes!
"Thy two breasts are like two young deer that are twins.
Thy neck is as a tower of ivory.
Thine eyes are like the fish pools by the white harbour." (1)
My passions are boundless. My passions are the same as yours.
Feel your passion. Let it fill you. Let it fill you with your power and beauty!

To Man:
You, Man of my heart. Do you remember me?
Do you remember the love we once shared?
I long for a mate, thirsty like water.
Come to me, I need your seed
or I wither and die, no fruit in my branches.
I want you, your potency, your fire.
Why do you resist me?
Is it because I am stronger than you in the darkness of the night?
You say I am an illusion, a dream.
I am your true dream.
I offer you rapture and peace for your solitary security.
"You deny and question, but mine eyes gleam on thee, lit with an alien light.
My lips proclaim mysteries.
My arms hold all that gods desire and fools reject. Behold me!" (3)

But I will not submit to you and be put beneath you like a serving maid.

I am his who dares to pay my price.

I ask too much, you say? Yet I give all. Why do you hold back?

I warn you. If you send me away, you will fall asleep,

And your rib will be taken out for your mate.

You banish me, but you will be cast out from the garden and struggle to be reborn.

I will return to remind you of what you really want.

How you feel broken with a mate who is but a part of you, like a crutch.

Her prophecy:

I hurt, I am ill. What is this place

where the children live in cement boxes and eat chemicals for food?

The land is burning with the raging fires of war

My trees are cut down, ruthlessly uprooted

Animals thanklessly consumed; my life blood spilled out for greed

I cannot live here. I must go . . .

"The flood plains of my heart will pour out over the land" (5)

And you will hear my voice in the howling winds and the hurricanes—

in the earthquakes and the volcanoes!

Her plea:

Awake, O thou whom my soul loveth, and come

Come into my garden, where the spices flow out.

"Come into your garden and eat the pleasant fruits." (1)

Come to the temple for the holy rites of love.

"Set me as a seal upon thine heart" (1)

Set your soul free.

Please, call me back.....

Sources: (1) "The Song of Songs, that is Solomon's," Bible, King James edition;

(2) Pagels, *The Gnostic Gospels*;

(3) Sterling, *Lilith: A Dramatic Poem;*

(4) *Inner Visions*, women's chants;

(5) Wolkstein and Kramer, *Inanna, Queen of Heaven and Earth*.

Appendix
Further Astronomical Notes

For the research in this book, I have used Black Moon data from *The New International Ephemerides 1950-2050* by Aureas under Editions St. Michel and as given on the Astrodienst website, www.astro.com. They both give the generally-used mean and true positions. A "corrected" position is available in *The Complete Ephemerides 1930-2030*, published later by Aureas. Even further mathematical research has been done by Francis Santoni in his *Ephemerides of the True Black Moon* (also available at aureas.org).

David Cochrane, author of the Kepler astrology program, reports that most astrology (and astronomy) software relies on the Swiss Ephemeris routines to produce calculations. The Swiss Ephemeris at the present time offers very accurate positions of True Lilith and Mean Lilith. However, there are complex technical issues in the determination of True Lilith and some people have suggested variations in the calculations.

The main problem in the determining the position of Lilith, Cochrane explains, is that the elliptical path of the Moon is constantly changing. Imagine a race car or a runner going around the track. Although the track is not quite an elliptical shape, this image will suffice for our purposes now. The race car stays in the same line so therefore it repeats its orbit over and over. However, in the case of the path of the Moon, it moves very far off the path and therefore behaves very differently from the race car. One of the main reasons why the Moon does not stay on a stable path is that it is pulled by the gravitational force of the Sun. This disturbance in the Moon's path is called perturbation. The planets also experience perturbations from the Sun and other planets but they do not change the path of the planet as much as the perturbations of the Sun change the path of the Moon.

The question then arises as to what the "real" elliptical path of the Moon is. There are several choices. One choice is to define a constantly changing elliptical path. An elliptical path can be drawn for the Moon at any point in time. Using this model, the elliptical path of the Moon is constantly changing very quickly. This is analogous to our race track shifting around rather quickly and by a substantial amount. In many ways, this is a reasonable decision. Why should we think there is one real path of the Moon that the Moon deviates from? We simply view the Moon as constantly on an elliptical path and that path is constantly changing. This is the definition generally used by astrologers.

Two other choices are similar but not identical. We can use an average position of the lunar apogee and this is the mean Lilith position. The other choice is to interpolate between successive positions of the lunar apogee when the Moon is conjunct the lunar apogee. This is similar to the mean Lilith position except that we may also interolate between times when the Moon is opposite the lunar apogee, i.e., in the direction of the lunar perigee and this results in an apogee and perigee positions (i.e., the Lilith position, and the point opposite theLilith position) that are not opposite to each other. This position is sometimes referred to as a natural Lilith position.

Riyal Program

Costa Rican Astrologer Juan Antonio Revilla gives numerous variations of the Lunar Apogee/Black Moon in his Riyal program including the Natural Lilith position. For more information and a full explanation of the mathematic technicalities, see his discussion at www.expreso.co.cr/centaurs/blackmoon/barycentric.html.

To summarize his key comments: Tracking the true Black Moon is a complex mathematical task due to the ever-changing gravitational dance between Sun-Moon-Earth. Continual variations in the Moon's orbital elements, such as speed, position and angle, defy precision. The Moon's orbit wobbles, pulled by the Sun, the planets and other solar systemic influences. The Earth also wobbles within itself around the Earth-Moon barycenter. It seems that any calculation cannot be totally correct, as results depend on the base point of data collection. There are three choices:

1. Topocentric, from the Earth's surface, which would obviously vary by latitude and longitude;

2. Geocentric, from the center of the Earth. The Moon does not actually orbit around the center of the Earth, though some calculations are made using this geocentric point; and

3. Barycentric, from the Earth-Moon barycenter. The gravitational center around which the Earth-Moon system spins, the barycenter, is located inside the Earth about one-third from Earth's center.

Published tables generally are geocentric-based. The "true" oscillating orbit is generally calculated on a specific, instantaneous time-space measurement of orbital elements. This data is then subjected to Kepler's Laws of Planetary Motion, the astronomical formula that calculates the hypothetical path of the object based on that data. Since the data continually changes and is not totally predictable, this "true" position really does not live up to its name. However it is widely used as a truer representation of the variant movement.

The Unicorn Connection

The Black Moon is often considered synonymous with the Lunar Apogee, the furthest point of Moon from Earth. The data is workably close, pointing to the same zodiacal point, yet is not precise. You can see in the astronomical graphic that they are different distances from Earth, therefore representing different perspectives. The Black Moon as second focus of the Moon's orbit is near the Earth, participating in the personality obfuscations of the Moon. The Lunar Apogee is the point at which the Moon is furthest from Earth, most free from Earth and most open to divine influence. Using the mean Black Moon gives us a "clean," archetypal position.

The pioneering French astrologer Jean Carteret brought the "black lights," the Black Moon and Black Sun, to the attention of the astrology world. In *La Lune Noir, un Vertige d'Absolu*, French astrologer Luc Bige opens further mythic dimensions of Black Moon Lilith by naming the five points on the long axis of the Moon's orbit.

1. The Lunar Perigee he calls Priapus, a phallic god of rustic origin. For the dark, primordial and raw sexuality of Lilith, he is a very appropriate partner.

2. Earth is our home ground. The modern scientific vision called the Gaia Hypothesis hearkens back to the ancient perception of Earth as a living goddess.

3. The barycenter, the center point of the ellipse located inside the Earth, is called Hecate. She is a Moon-related goddess who stands at a triple crossroads as a way-shower.

4. True Black Moon, the second, empty focus of the ellipse, is identified as the ever-elusive and shifty Lilith.

5. The Lunar Apogee, or mean Black Moon, is seen as the Unicorn. The image of the Unicorn symbolizes the purity and soul authenticity of the Lunar Apogee, which Bige describes as a sparkling divine radiance, a place of grace, a point of telepathic contact with extra-planetary energies. Located on the same axis, one reaches the Lunar Apogee radiance only through the jittery, shimmering veil of the "true" Lilith focal point.

I hope you get the sense that technical complexities in the mathematics are inherent to Lilith's mysteries. Don't get bogged down trying to understand it all! Use whatever positions of Black Moon you can find and see what you discover as you enter the labyrinth.

Chart Data

Data has been checked through the AstroDataBank program as much as possible. Otherwise I have used the best possible online sources.

.ADB = AstroDataBase, with Rodden Rating

> AA= accurate, data officially recorded
>
> A = accurate, data from reliable source
>
> B= data from biography
>
> C= caution, source unreliable
>
> DD= dirty data, conflicting information
>
> X= data with no birth time. Except for Ascendant and house information, time is not

necessary for determining mean Black Moon. The position of true Black Moon moves quickly, so birth time is more critical.

Aguilera, Christina, December 18, 1980, Staten Island, NY, 10:46 a.m., ADB#A

Aoki, Hiraoki "Rocky," October 9, 1938, Tokyo, Japan, 12:45 p.m., ADB#A

Albright, Madeleine Korbel, nee Marie Jana Korbelova, May 15, 1937, Prague, time unknown, ADB#X

Allen, Woody, December 1, 1935, Bronx, NY, 10:55 p.m., ADB#AA

Amory, Cleveland, September 2, 1917, Nahant, MA, 6:30 a.m., ADB#A

Amos, Tori, August 22, 1963, Newton, NC, 1:10 p.m., ADB#AA

Amundsen, Roald, July 16, 1872, Sarpsborg, Norway, 5:00 a.m., ADB#C

Andersen, Hans Christian, April 2, 1805, Odense, Denmark, 1:00 a.m., ADB#AA

Andrews, Dame Julie, October 1, 1935, Walton on Thames, England, 6:00 a.m., ADB#B

Angelou, Maya, April 4, 1928, St. Louis, MO, 2:10 p.m., ADB#AA

Ann-Margret, April 28, 1941, Valsjobyn, Sweden, 4:30 a.m., ADB#A

Antoinette, Marie, November 2, 1755, Vienna, Austria, 7:30pm LMT, ADB#A

Armani, Giorgio, July 11, 1934, Piacenza, Italy, 7:20 a.m., ADB#AA

Armstrong, Lance, September 18, 1971, Plano, TX, time unknown, ADB#X

Astor, Mary, nee Lucille Langhanke, May 3, 1906, Quincy, IL, 8:10 p.m., ADB#A

Auden, Wystan Hugh, February 21, 1907, York, UK, 6:45 a.m., ADB#DD date confirmed

Badu, Erykah, nee Erica Abi Johnson, February 26, 1971, South Dallas, TX, time unknown, numerous web sources suggest general agreement on birth data

Bailey, Alice, June 16, 1880, Manchester, England, 7:32 a.m. GMT, ADB#C source unknown

Baker, Josephine, June 3, 1906, St. Louis, MO, 11:00 a.m., ADB#B

Barr, Candy, nee Juanita Dale Slusher, July 6, 1935, Edna, TX, 4:00 a.m., ADB#AA

Barrymore, Drew, February 22, 1975, Culver City, CA, 11:51 a.m., ADB#AA

Beardsley, Aubrey, August 21, 1872, Brighton, England, 5:18 p.m. GMT, ADB#C, source unknown

Beatles, Asteroid 8749 discovery date, April 3, 1998, Gold Coast, Australia, no time given, NASA's Jet Propulsion Laboratory Small-Body Database Browser, //ssd.jpl.nasa. gov/sbdb.cgi

Bergman, Ingmar, July 14, 1918, Uppsala, Sweden, 1:00pm CET, ADB#C (source unknown)

Bernanos, Georges, February 20, 1888, Paris, 9:00 a.m., ADB#AA

Besant, Annie, October 1, 1847, London, England, 5:29 p.m., ADB#A

Birkin, Jane, December 14, 1946, London, England, 11:00 a.m., ADB#C, conflicting times

Bjork, November 21, 1965, Reykyavik, Iceland, 7:50am WAT, ADB#A

Black, Shirley Temple, April 23, 1928, Santa Monica, CA, 9:00 p.m., ADB#AA

Blaine, David, ne David Blaine White, April 4, 1973, Brooklyn, NY, unknown time, numerous web sources suggest general agreement on birth data

Blegen, Judith, April 27, 1941, Lexington, Kentucky, 4:03 a.m., ADB#AA

Bobbitt, John Wayne, March 23, 1967, Buffalo, NY, 10:00 a.m., ADB#A

Bon Jovi, Jon, ne John Bongiovi, March 2, 1962, Perth Amboy, NJ, 8:45 p.m., ADB#AA

Bonaparte, Napoleon, August 15, 1769, Ajaccio, Corsica, 11:30 a.m. LMT, ADB#A

Bono, ne Paul Hewson, May 10, 1960, Dublin, Ireland, 2:00am GDT, ADB#A

Brandt, Willie, ne Karl H. Frahm, September 18, 1913, Lubeck, Germany, 12:45 p.m., ADB#AA

Brecht, Bertold, February 10, 1898, Augsburg, Germany, 4:30 a.m., ADB#AA

Brejnev, Leonid, December 19, 1906 (Julian calendar), Komenskoie, Ukraine, time unknown

Brooks, Garth, February 7, 1962, Tulsa, OK, 1:07 p.m., ADB#A

Bundy, Ted, November 24, 1946, Burlington, VT, 10:35 p.m., ADB#AA

Bush, George Herbert Walker, June 12, 1924, Milton, MA, 11:45 a.m., ADB#A

Byron, George Gordon, Lord, January 22, 1788, London, England, 2:00 p.m., ADB#AA

Cardin, Pierre, July 2, 1922, San Biagio Di Callalta, Italy, 2:00 p.m. MET, ADB#AA

Carrey, Jim, January 17, 1962, Newmarket, ON, Canada, 2:30 a.m. EST, ADB#A

Carl XV of Sweden, May 3, 1826, Stockholm, Sweden, 2:15 p.m., ADB#AA

Cash, Johnny, February 26, 1932, Kingsland, AR, 7:30 a.m. CST, ADB#A

Catherine II, "The Great," May 2, 1729, Stettin, Germany, 2:30 a.m., ADB#B

Cayce, Edgar, March 18, 1877, Hopkinsville, KY, 2:03 p.m., ADB#C, time unverified

Chamberlain, Richard, March 31, 1934, Los Angeles, CA, 6:20 p.m., ADB#AA

Chavez, Hugo, July 28, 1954, Sabaneta, Venezuela, 3:30-4:00 a.m.?, ADB#DD, reports on time vary, reports on time vary

Cher, nee Cherilyn LaPiere, May 20, 1946, El Centro, CA, 7:25 a.m. PST, ADB#A

Chiba, Sonny, ne Sadao Maeda, January 23, 1939, Fukuoka, Japan, time uknown, numerous web sources suggest general agreement on birth data

Child, Julia, August 15, 1912, Pasadena, CA, 11:30 p.m. PST, ADB#AA

Chomsky, Noam, December 7, 1928, Philadelphia, PA, 7:00 a.m., ADB#A

Chopin, Frederik, March 1, 1810, Zelazowa Wola, Poland, 6:00 p.m., ADB#DD unverified time

Chopra, Deepak, October 22, 1946, New Delhi, India, 3:45 p.m. IST, ADB#A

Christie, Agatha, September 15, 1890, Torquay, England, 4:00 a.m. GMT, ADB#A

Clapton, Conor, August 21, 1986, London, England, 6:20 a.m., ADB#A. Death of, March 20, 1991, New York, New York, 11:05 a.m., www.eric-clapton.co.uk/interviewsandarticl es/loryinterview.htm and www.obituariestoday.com/Obituaries/ObitShow.php?Obitu ary_ID=29613/

Clapton, Eric, March 30, 1945, Ripley, England, 8:45 p.m.?, ADB#DD conflicting times

Clarkson, Kelly, April 24, 1982, Forth Worth, TX, time unknown, numerous web sources suggest general agreement on birth data

Cohen, Leonard, September 21, 1934, Montreal, Canada, 6:45 a.m. EDT, ADB#A

Comaneci, Nadia, November 12, 1961, Gheorghe Dej, Romania, 10:00 a.m. EET, ADB#A

Connors, Jimmy, September 2, 1952, E. St. Louis, MO, 10:30 a.m., ADB#AA

Cooper, Alice, ne Vincent Furnier, February 4, 1948, Detroit, MI, 10:33 p.m., ADB#AA

Copernicus, Nicolaus, February 28, 1473, Torun, Poland, 5:13 p.m. LMT, ADB#AA

Coppola, Sofia, May 14, 1971, New York City, time unknown, numerous web sources suggest general agreement on birth data

Crow, Sheryl, February 11, 1962, Kennett, MO, 9:58 a.m., Astrotheme

Curie, Marie, November 7, 1867, Warsaw, Poland, 10:36 a.m., ADB#AA

Dalai Lama, His Holiness, the 14th,Tenzin Gyatso, July 6, 1935, Taktser, Tibet, 4:38 a.m.?, ADB#A. Although ADB gives an A rating to a 4:38 a.m. birth time, it reports various contradictions. For our purposes, not time-dependent, we use dawn as a common fallback position, which is most likely close. Birthdate given on www.dalailama,com

Davis, Miles, May 26, 1926, Alton, IL, 5:00 a.m., ADB#AA

DeHaviland, Olivia, July 1, 1916, Tokyo, Japan, 10:30 a.m., ADB#A

DeLors, Jacques, July 20, 1925, Paris, France, 12:30 p.m., Astrotheme, http://encyclopedia. stateuniversity.com and http://ec.europa.eu/commission

DeNiro, Robert, August 17, 1943, Brooklyn, NY, 3:00 a.m., ADB#A

Depp, Johnny, June 9, 1963, Owensboro, KY, 8:44 a.m., ADB#AA

Derrida, Jacques, July 15, 1930, El Biar, Algeria, 8:30 a.m., ADB#AA

Descartes, Rene, March 31, 1596, Descartes, France, time unknown, ADB#X, no time

Dickens, Charles, February 7, 1812, Portsmouth, England, 7:50 p.m., ADB#A

Dickinson, Emily, December 10, 1830, Amherst, MA, 4:40 a.m. LMT, ADB#B

Dionne Quintuplets, May 28 1934, Callander, Canada, 3:56 a.m., born within a few minutes, ADB#A

Disney, Walt, December 5, 1901, Chicago, IL, 12:35 a.m. CST, ADB#A

Doenitz, Karl, September 16 1891, Grunau-bei-Berlin, Germany, 3:30 p.m., ADB#AA

Dore, Gustave, January 6, 1832, Strasbourg, France, 6:00 a.m., ADB#AA

Douglas, Kirk, ne Issur Demsky, December 9, 1916, Amsterdam, NY, 10:15 a.m., ADB#AA

Dreyfus, Alfred, October 9, 1859, Mulhouse, France, 3:00 p.m., ADB#AA

DuMaurier, Daphne, May 13, 1907, London, 4:00 p.m.?, time uncertain, ADB#DD

Dunst, Kirsten, April 30, 1982, Point Pleasant, NJ, time unknown, numerous web sources suggest general agreement on birth data

Eberhardt, Isabelle, February 17, 1877, Geneva, Switzerland, 9:30 a.m., ADB#AA

Eco, Umberto, January 5, 1932, Alessandria, Italy, 6:30 p.m. MET, ADB#AA

Einstein, Albert, March 14, 1879, Ulm, Germany 11:30 a.m. LMT, ADB#AA

Eliade, Mircea, March 13, 1907, Bucharest, Romania, 5:00 a.m., ADB#A

Elizabeth I, Queen of England, September 17 1533, Greenwich, England, 2:54 p.m. LMT, ADB#AA

Estefan, Gloria, September 1, 1957, Havana, Cuba, unknown time, numerous web sources suggest general agreement on birth data, including her website, www.gloriaestefan.com

Etheridge, Melissa, May 29, 1961, Leavenworth, KS, 1:15 p.m., ADB#B

Eugenie, Empress, May 5, 1826, Granada, Spain, unknown time, ADB#C

Faltskog, Agneta, April 5, 1950, Jonkoping, Sweden, unknown time, numerous web sources suggest general agreement on birth data

Farmer, Kelle Marie, October 1, 1980, Cardiff, Wales, time unknown, numerous web sources suggest general agreement on birth data

Farrow, Mia, February 9, 1945, Los Angeles, CA, 11:27 a.m., ADB#AA

Federer, Roger, August 8, 1981, Basel, Switzerland, 8:40 a.m., www.rogerfederer.com/en/rogers/profile/index.cfm

Ferrer, David, April 2, 1982, Javia, Spain, time unknown, numerous web sources suggest general agreement on birth data, including official ATP website

Fischer, Bobby, March 9, 1943, Chicago, IL, 2:39 p.m. CWT, ADB#B

Foster, Jodie, November 19, 1962, Los Angeles, CA, 8:14 a.m. PST, ADB#AA

Fox, Michael J., June 9, 1961, Edmonton, Canada, 12:51 a.m., ADB#A

Frost, Robert, March 26, 1874, San Francisco, CA, 7:00 a.m.?, ADB#DD, time in question

Fugate, Caril Ann, July 30, 1943, Lincoln, Nebraska, 12:55 p.m., ADB#AA

Gabriel, Peter, February 13, 1950, Woking, England, 4:30 p.m., ADB#A

Gacy, John Wayne, March 17, 1942, Chicago, IL, 12:29 a.m. CWT, ADB#AA

Galileo, Gallilei, February 15, 1564, 3:31 p.m. LMT, ADB#A

Gandhi, Mohandas, October 2, 1869, 7:11 a.m. LMT, Porbandar, India, ADB#A

Gandhi, Sonia, December 9, 1946, Lusiana, Italy, 9:30 p.m., ADB#AA

Garbo, Greta, ne Gustafsson, September 18, 1905, Stockholm, Sweden, 7:30 p.m., ADB#AA

Garfunkel, Art, November 5, 1941, New York, NY, 11:00 p.m., ADB#DD

Garland, Judy, June 10, 1022, Grand Rapids, MN, 6:00 a.m. CST, ADB#AA

Gates, Bill, October 28, 1955, Seattle, WA, 10:00 p.m. PST, ADB#A

Geldof, Bob, October 5, 1951, Dublin, Ireland, 2:20 p.m., ADB#A

Gibson, Mel, January 3, 1956, Peekskill, NY, 4:45 p.m., ADB#B

Giovanni, Nikki, June 7, 1943, Knoxville, TN, 6:45 a.m., ADB#AA

Godel, Kurt, April 28, 1906, Brun, Moravia, time unknown, nNumerous academic web sources suggest general agreement on birth data including citations from birth certificate in private papers held in Princeton University Library

Goethe, Johann Wolfgang von, August 28, 1749, Frankfurt-am-Main, Germany, 12:00 p.m., ADB#B

Goodman, Linda, April 10, 1925, Morgantown, WV, 1:56 p.m., ADB#A

Gorbachev, Mikhail, March 2, 1931, Privolnoye, Russia, 8:30 p.m.?, ADB#DD, time in question

Gosling, Ryan, November 12, 1980, London, ON, Canada, time unknown, numerous web sources suggest general agreement on birth data

Goya, Francisco de, March 30, 1746, Fuentetodos, Spain, no time, ADB#X, numerous web sources suggest general agreement on birth data

Grant, Cary, ne Archibald Leach, January 18, 1904, Bristol, England, 1:07 a.m. GMT, ADB#B

Grant, Joan, April 12, 1907, London, England, 8:00 a.m., ADB#A

Griffin, Merv, July 6, 1925, San Mateo, CA, 4:45 a.m., ADB#AA

Griffith, Andy, June 1, 1926, Mount Airy, NC, 3:00 p.m., ADB#AA

Guinness, Sir Alec, April 2, 1914, London, 5:45 a.m., ADB#C

Guthrie, Woody, July 14, 1912, Okemah, OK, time unknown, numerous web sources suggest general agreement on birth data, including www.woodyguthrie.org/biography

Gyllenhaal, Jake, December 19, 1980, Los Angeles, CA, 8:08 a.m., ADB#AA

Hamill, Mark, September 25, 1951, Oakland, CA, 2:43 p.m., ADB#AA

Harrer, Heinrich, July 6, 1912, Huttenberg, Austria, time unknown, numerous web sources suggest general agreement on birth data, including obituary in the U.K. Telegraph

Harrison, George, February 24, 1943, Liverpool, England, 11:42 p.m., ADB#A

Harrison, Asteroid 4149 discovery date, March 9, 1984, Flagstaff, AZ, time not given, NASA's Jet Propulsion Laboratory Small-Body Database Browser, //ssd.jpl.nasa.gov/sbdb.cgi

Hefner, Hugh, April 9, 1926, Chicago, IL, 4:20 p.m., ADB#AA

Hendrix, Jimi, November 27, 1942, Seattle, WA, 10:15 a.m., ADB#AA

Henin, Justine, June 1, 1982, Liege, Belgium, 4:30 p.m., Astrotheme

Hepburn, Katherine, May 12, 1907, Hartford, CT, 5:47 p.m., ADB#AA

Hindenberg, Paul von, October 2, 1847, Poznan, Poland, 3:00 p.m., ADB#AA

Hingis, Martina, September 30, 1980, Kosice, Slovakia, time unknown, ADB#X

Hitchcock, Alfred, August 13, 1899, London, England, 3:15 a.m.?, ADB#DD, time in question

Hoffa, Jimmy, February 14, 1913, Brazil, IN, 6:52 a.m., ADB#AA

Honda, Soichiro, November 17, 1906, Komyo Village (now Tenryu City), Japan, time unknown, worldhonda.com/history

Hopkins, Anthony, December 31, 1937, Port Talbot, Wales, 9:15 a.m., ADB#A

Horne, Marilyn, January 16, 1934, Bradford, PA, 3:20 p.m., ADB#AA

Houston, Whitney, August 9, 1963, Newark, NJ, 8:55 p.m., ADB#AA

Hume, David, May 7, 1711, Edinburgh, Scotland, time unknown, numerous web sources suggest general agreement on birth data

Huston, Anjelica, July 8, 1951, Beverly Hills, CA, 6:29 p.m., ADB#AA

Hynde, Chrissie, September 7, 1951, Akron, OH, 10:20 a.m., ADB#AA

Irwin, Steve, February 22, 1962, Essendon, Australia, 4:18 a.m.? AEST, ADB#DD, time unverified

Jackman, Hugh, October 12, 1968, Sydney, Australia, unknown time, numerous web sources suggest general agreement on birth data, includingwww.hugh-jackman.com/hugh-information/biography, www.hugh-jackman.net/hugh-jackman-biography.aspx

Jackson, Jesse, October 8 1941, Greenville, SC, 2:15 p.m. EST, ADB#A

Jagger, Elizabeth Scarlett, March 2, 1984, New York City, 1:37 a.m., ADB#A

Jagger, Mick, July 26, 1943, Dartford, England, 2:30 a.m., ADB#A

Jean, Grand Duke of Luxembourg, January 5, 1921, Luxembourg, time unknown, www. gouvernement.lu/dossiers/famille_grand_ducale/chregneuk/infobase/cvjean.html, official website of the Luxembourg government

Jones, Nella, May 4, 1932, Belvedere Marshes, England, 10:30 a.m., ADB#A

Joplin, Janis, January 19, 1943, Port Arthur, TX, 9:45 a.m. CWT, ADB#AA

Josephine-Charlotte, October 11, 1927, Brussels, Belgium, 7:50 a.m., ADB#AA, www.gou vernement.lu/dossiers/famille_grand_ducale/chregneuk/infobase/cvjosephine.html

Kahlo, Frida, July 6, 1907, Coyoacan, Mexico, 8:30 a.m., ADB#AA

Kaysen, Susanna, November 11, 1948, Boston, MA, 6:05 a.m., ADB#AA

Keaton, Diane, January 5, 1946, Los Angeles, CA, 2:49 a.m., ADB#AA

Keeler, Christine, February 22, 1942, London, England, 11:15 a.m., ADB#A

Keller, Helen, June 27, 1880, Tuscumbia, AL, 4:02 p.m.?, ADB#DD, time unconfirmed

Kennedy, Edward "Ted," February 22, 1932, Dorchester, MA, 3:58 a.m. EST, ADB#AA

Kennedy, Joan, September 5, 1935, New York City, 6:10 a.m., ADB#AA

Kennedy, John F., May 29 1917, Brookline, MA, 3:00 p.m. EST, ADB#A

Kennedy, Rose, July 22, 1890, Boston, 10:00 p.m.?, ADB#DD

Kerouac, Jack, March 12, 1922, Lowell, MA, 5:00 p.m., ADB#B

Kesey, Ken, September 17, 1935, La Junta, CO, 4:24 p.m., ADB#AA

Kiki de Montparnasse, nee Alice Prin, October 2, 1901, Chatillon-sur-Seine, France, 8:51 a.m., ADB#AA

King, Martin Luther, Jr., January 15, 1929, Atlanta, GA, 12:00 p.m., ADB#A

Kingsley, Ben, ne Krishna Bhanji, December 31, 1943, Snainton, Scarborough, Yorkshire, UK, time unknown, numerous web sources suggest general agreement on birth data

Kinnear, Greg, June 17, 1963, Logansport, IN, 3:21 p.m., ADB#AA

Kovalevskaya, Sofia, January 15, 1850, Moscow, Russia, time unknown, numerous academic websites cite birth data as given, including biographical information in *Contemporary Mathetmatics: The Legacy of Sonya Kovalevskaya*, proceedings of a Symposium Sponsored by The Association for Women in Mathematics and The Mary Ingraham Bunting Institte held October 25-28, 1985, published by the American Mathematical Society. Volume 64

Krauss, Alison, July 23 1971, Decatur, IL, 7:48 p.m., ADB#AA

Kudrow, Lisa, July 30, 1963, Encino, CA, 4:37 a.m., ADB#AA

Kushi, Michio, May 17, 1926, Kokawa, Japan, 2:00 a.m., ADB#A

Lamborghini, Ferruccio, April 28, 1916, Cento Fe, Italy, 5:00 or 5:30 p.m., ADB#AA

Landon, Michael, October 31, 1936, Jamaica, NY, 12:12 p.m., ADB#AA

lang, k.d., November 2, 1961, Edmonton, Canada, 2:03 a.m., ADB#AA

Latifa, Queen, March 18, 1970, Newark, NJ, 8:02 a.m., ADB#AA

Law, Jude, December 29, 1972, Lewisham, England, 6:00 a.m., Astrotheme, numerous web sources suggest general agreement on birth date and place, time is given on www.vegaattractions.com/celebrity/stars/lawjude.html

Lawrence, David Herbert, September 11, 1895, Eastwood, UK, 9:45 a.m., ADB#A

Lennon, John, October 9, 1940, Liverpool, England, 6:30 p.m., ADB#A

Lennon, Asteroid 4147 discovery date, January 12, 1983, Flagstaff, AZ, time not given, NASA's Jet Propulsion Laboratory Small-Body Database Browser, ssd.jpl.nasa.gov/sbdb.cgi

Lewinsky, Monica, July 23, 1973, San Francisco, CA, 12:21 p.m., ADB#AA

Lewis, Jerry Lee, September 29, 1935, Ferriday, LA, 3:00 p.m., ADB#AA

Lindberg, Charles, February 4, 1902, Detroit, MI, 1:30 a.m., ADB#AA

Luxembourg Royal Marriage, April 9, 1953, Luxembourg, time unknown, official website of the Luxembourg government, www.gouvernement.lu/dossiers/famille_grand_ducale/chregneuk/infobase/cvjean.html

Lynn, Loretta, April 14, 1932, Butcher Hollow (Van Lear), KY, 4:00 p.m. EST, ADB#C, unknown source

Macchio, Ralph, November 4, 1961, Huntington, NY, time unknown, numerous web sources suggest general agreement on birth data

MacLaine, Shirley, April 24, 1934, Richmond, VA, 3:57 p.m., ADB#AA

Maharishi Mahesh Yogi, January 12, 1917, Jabalpur, India, time unknown, ADB#X

Mako, Princess of Japan, Octoeber 23, 1991, Tokyo, Japan, 11:41 p.m., ADB#AA

Malcolm X, ne Malcolm Little, May 19, 1925, Omaha, Nebraska, 12:20 p.m., ADB#AA

Mandela, Nelson, July 18, 1918, Tanskrei, South Africa, time unknown, nobelprize.org/nobel_prizes/peace/laureates/1993/mandela-bio.html

Manning, Peyton, March 24, 1976, New Orleans, LA, time unknown, www.peytonmanning.com, www.nfl.com/players/peytonmanning/profile?id=MAN51 5097

Manson, Charles, November 12, 1934, Cincinnati, OH, 4:40 p.m. EST, ADB#AA

Marley, Bob, February 6, 1945, St. Ann's Bay, Jamaica, 2:30 a.m., ADB#XX, date appears firm

Martin, Ricky, December 24, 1971, Hato Rey, Puerto Rico, 5:00 p.m., ADB#AA

Mathis, Johnny, September 30, 1935, Gilmer, TX, 12:00 a.m., ADB#AA

Matsuda, Seiko, ne Kamachi Noriko, March 10, 1962, Fukuoda, Japan, 5:30 p.m., ADB#A

McCaffrey, Anne, April 1, 1926, Cambridge, MA, 1:30 p.m., ADB#AA

McCartney, Sir Paul, June 18, 1942, Liverpool, England, 2:00 p.m., ADB#A

McCartney, Asteroid 4148 discovery date, July 11, 1983, Flagstaff, AZ, time not given, NASA's Jet Propulsion Laboratory Small-Body Database Browser, ssd.jpl.nasa.gov/sbdb.cgi

McGuire, Phyllis, February 14, 1931, Middletown, OH, 8:25 a.m., ADB#AA

McVeigh, Timothy, April 23, 1968, Lockport, NY, 8:19 a.m., ADB#AA

McVeigh Execution, June 11, 2001, Terre Haute, IN, 7:14 a.m., various news archives give this time, including CNN and BBC

Menuhin, Yehudi, April 22, 1916, New York, NY, 11:30 p.m., ADB#B

Mistinguett, ne Jeanne Florentine Bourgeois, April 3, 1875, Enghien, France, 10:00 p.m., ADB#AA

Mitchell, Joni, November 7, 1943, Fort MadLeod, Canada, 10:00 p.m., ADB#A

Mitterand, Francois, October 26, 1916, Jarnac, France, 4:00 a.m. GMT, ADB#AA

Monroe, Marilyn, ne Norma Jean Mortenson, June 1, 1926, Los Angeles, CA, 9:30 a.m., ADB#AA

Moore, Michael, April 23, 1954, Flint, MI, 12:45 p.m., ADB#AA

Moreno, Rita, December 11, 1931, Humacao, Puerto Rico, time unknown, numerous web sources suggest general agreement on birth data

Mumba, Samantha, January 18, 1983, Dublin, Ireland, 12:00 p.m.?, Astrotheme, numerous web sources suggest general agreement on birth data

Myss, Caroline, December 2, 1952, Chicago, IL, 8:00 a.m., ADB#A

Nadal, Rafael, June 3, 1986, Manacor, Spain, time unknown, numerous web sources suggest general agreement on birth data

Nalbandian, David, January 1, 1982, Cordoba, Argentina, time unknown, numerous web sources suggest general agreement on birth data, including official ATP website

NASA, (U.S.) National Aeronautic and Space Administration, October 1, 1958, 8:00 a.m., Hampton, VA, ADB#A

Nietzsche, Friedrich, October 15, 1844, Rocken, Germany, 10:00 a.m., ADB#B

Nin, Anais, February 21, 1903, Neuilly-sur-Seine, France, 8:25 a.m., correspondence to author

Neumann, Therese, April 8 or 9, 1898, Konnersreuth, Rumania, 11:15 p.m. or 12:15 a.m., one hour time difference, ADB#DD

Newman, Paul, January 26, 1925, Cleveland, OH, 6:30 a.m., ADB#AA

Novarupta Volcano eruption, June 6, 1912, Kodiac, AK, daytime, historic record, exact start unknown

Obama, Barack, August 4, 1961, Honolulu, HI, 7:24 p.m., birth certificate published on web by his campaign to prove citizenship, June 2008

Oklahoma Bombing, April 19, 1995, Oklahoma City, OK, 9:02 a.m., ADB#A, www.oklahomacitynationalmemorial.org

Oldman, Gary, March 21, 1958, London, 10:56 a.m., Astrotheme

Olivier, Sir Lawrence, May 22, 1907, Dorking, UK, 5:00 a.m., ADB#AA

Onassis, Jacqueline Bouvier Kennedy, July 28, 1929, Southampton, NY, 2:30 p.m. EDT, ADB#AA

O'Neal, Ryan, April 20, 1941, Los Angeles, CA, 9:34 a.m., ADB#A

O'Neal, Tatum, November 5, 1963, Los Angeles, CA, 3:38 a.m., ADB#AA

Palin, Michael, May 5, 1943, Sheffield, England, 11:45 a.m. GDWT, ADB#A

Parks, Rosa, February 4, 1913, Tuskegee, AL, time unknown, ADB#X

Pavarotti, Luciano, October 12, 1935, Modena, Italy, 1:40 a.m., ADB#AA

Peck, Gregory, April 5, 1916, LaJolla, CA, 8:00 a.m., ADB#AA

Pennetta, Flavia, February 25, 1982, Brindisi, Italy, time unknown, official WTA website

Petursson, Johann, February 9, 1913, Dalvik, Iceland, 12:30 p.m.?, Astrotheme

Picasso, Pablo, October 25, 1881, Malaga, Spain, 11:15 p.m., ADB#AA

Pickford, Mary, nee Gladys Smith, April 8, 1892, Toronto, Canada, 3:00 a.m.?, ADB#DD unverified time

Pieck, Anton, April 19, 1895, Den Helder, Netherlands, 2:10 a.m., ADB#AA

Pitt, Brad, December 18, 1963, Shawnee, OK, 6:31 a.m., ADB#A

Plant, Robert, August 20, 1948, West Bromwich, England, 3:05 p.m., ADB#C, unknown source

Pollack, Sydney, July 1, 1934, Lafayette, IN, 4:30 p.m., ADB#AA

Pope Benoit XIV, ne Prospero Lambertini, March 31, 1675, Bologna, Italy, time unknown, New Advent Catholic Encyclopedia http://www.newadvent.org

Portman, Natalie, June 9, 1981, Jerusalem, Israel, time unknown, numerous web sources suggest general agreement on birth data

Rabin, Yitzhak, March 1, 1922, Jerusalem, Israel, 3:50 p.m.,?, ADB#C, source uncertain

Radical Gay Faerie Gathering (first), August 31, 1979, (desert west of) Tucson, AZ, 6:00 p.m., astrologer Hidea report on YouTube, September 2008, for time; place corrected in personal correspondence with participant

Rajneesh, Bhagwan Shree, December 11, 1931, Kutchwada, India, 5:13 p.m. INT, ADB#A

Ravel, Maurice, March 7, 1875, Ciboure, France, 10:00 p.m., ADB#AA

Reeve, Christopher, September 25, 1952, Manhattan, NY, 3:12 a.m., ADB#A

Renoir, Auguste, February 25, 1841, Limoges, France, 6:00 a.m., ADB#AA

Richard I, the Lion-Hearted, September 15, 1157, Oxford, England, 3:00 a.m.?, ADB#C, various times given

Rilke, Rainer Maria, December 3, 1875, Prague, Czech Republic, 11:50 p.m., ADB#A. Numerous web sources give December 4 as birth date. Rodden's Astro-Data II, p.359, Astrolog #34 quotes Marian Gluntz, data from mother of "just before midnight," suggesting confusion across the midnight hour.

Rimbaud, Arthur, October 20, 1854, Charleville-Mezieres, France, 6:00 a.m., ADB#AA

Robredo, Tommy, May 1, 1982, Hostalric, Spain, time unknown, ATP official website

Roosevelt, Eleanor, October 11, 1884, New York, NY, 11:00 a.m. LMT, ADB#AA

Roosevelt, Franklin Delano, January 30, 1882, Hyde Park, NY, 8:45 p.m. LMT, ADB#AA

Rossellini, Isabella, June 18, 1952, Rome, Italy, 6:07 p.m., ADB#AA

Rossellini, Roberto, May 8, 1906, Rome, Italy, 12:50 p.m., ADB#AA

Rowling, J.K., July 31, 1965, Chipping Sodbury, England (near Bristol), time unknown, ADB#X, no time, birth date and place discussed on ADB as reliable

Roy, Arundhati, November 24, 1961, Shillong, India, time unknown, numerous web sources suggest general agreement on birth data

Rush, Geoffrey, July 6, 1951, Toowoomba, Australia, time unknown, numerous web sources suggest general agreement on birth data

Russell, Bertrand, May 18, 1872, Trellek, Wales, 5:45 p.m. GMT, ADB#B

Ryan, Meg, November 19, 1961, Fairfield, CT, 10:36 a.m., ADB#A

Ryder, Winona, Octoeber 29, 1971, Rochester, MN, 11:00 a.m., ADB#AA

Sagan, Francoise, nee Francoise Quoirez, June 21, 1935, Cajarc, France, 11:00 a.m., ADB#AA

Sampras, Pete, August 12, 1971, Washington, DC, time unknown, ADB#X

Selleck, Tom, January 29, 1945, Detroit, MI, 8:22 a.m., ADB#AA

Sciascia, Leonardo, January 8, 1921, Racalmuto, Italy, 11:45 a.m., ADB#AA

Shariff, Omar, April 10, 1932, Alexandria, Egypt, 5:30 p.m., ADB#B

Siffredi, Rocco, ne Rocco Tano, May 4, 1964, Ortona Mare, Italy, 11:30 a.m. MET, ADB#AA

Simmons, Gene, ne Klein, August 25, 1949, Haifa, Israel, 8:55 p.m., ADB#A

Simon, Paul, October, 13, 1941, Newark, NJ, 2:33 a.m., ADB#AA

Sinatra, Frank, December 12, 1915, Hoboken, NJ, 3:00 a.m. EST, ADB#A

Snoop Doggie Dog, October 20, 1971, Long Beach, CA, time uknown, ADB#X

Spears, Britney, December 2, 1981, McComb, MS, 1:30 a.m., ADB#A

Spencer, Lady Diana, Princess of Wales, July 1, 1961, Sandringham, England, 7:45 a.m., ADB#A

Spielberg, Steven, December 18, 1946, Cincinnati, OH, 6:16 p.m., ADB#AA

Starr, Ringo, ne Richard Starkey, July 7, 1940, Liverpool, England, 12:05 a.m., ADB#A

Starr, Asteroid 4150 discovery date, August 31, 1984, Flagstaff, AZ , time not given, NASA's Jet Propulsion Laboratory Small-Body Database Browser, //ssd.jpl.nasa.gov/sbdb.cgi

Stevenson, Robert Louis, November 13, 1850, Edinburgh, Scotland, 1:30 p.m., ADB#AA

Stewart, Martha, August 3, 1941, Jersey City NJ, 1:33 p.m., ADB#AA

Sting, ne Gordon Matthew Sumner, October 2, 1952, Wallsend, UK, 1:30 a.m., ADB#A

Streep, Meryl, June 22, 1949, Summit, NJ, 8:05 a.m. EDT, ADB#AA

Sutherland, Donald, July 17, 1935, St. John, Canada, 11:30 a.m., ADB#A

Swayze, Patrick, August 18, 1952, Houston, TX, 8:10 a.m., ADB#A

Swedenborg, Emanuel, February 8, 1689, Stockholm, Sweden, 6:48 a.m., ADB#XX. Swedish calendar changes put date in question, but most researchers agree on this date.

Tagore, Rabindranath, May 7 , 1861, Calcutta, India, 4:02 a.m.?, ADB#C, time uncertain

Tate, Sharon, January 24, 1943, Dallas, TX, 5:47 p.m., ADB#AA

Taylor, Elizabeth, February 27, 1932, London, England, 2:15 a.m., ADB#A

Teilhard de Chardin, Pierre, May 1, 1881, Orcines, France, 7:00 a.m., ADB#AA

Tiny Tim, ne Herbert Khaury, April 12, 1932, Manhattan, NY, time unknown, ADB#X no time

Tolkien, J. R. R., January 3, 1892, Bloemfontein, South Africa, 10:00 p.m.?, ADB#C, time in question

Travolta, John, February 18, 1954, Englewood, NJ, 2:53 p.m., ADB#AA

Updike, John, March 18, 1932, West Reading, PA, 3:45 p.m. EST, ADB#AA

Valadon, Suzanne, September 23, 1865, Bessines, France, 6:00 a.m., ADB#AA

Van Der Post, Sir Laurens, December 13, 1906, Philippolis, South Africa, 2:00 a.m.?, ADB#C, conflicting time

Van Gogh, Vincent, March, 30, 1853, Groot Zundert, Netherlands, 11:00 a.m., ADB#AA

Vedder, Eddie, nee Edward Seversen III, December 23, 1964, Evanston, IL, time unknown, numerous web sources suggest general agreement on birth data

Vuitton, Louis Moet Hennessey, August 4, 1821, Anchay, France, 3:00 a.m., ADB#AA

Walesa, Lech, September 29, 1943, Popow, Poland, 3:30 a.m., ADB#B

Warhol, Andy, August 6, 1928, Pittsburgh, PA, 6:30 a.m., ADB#B, time in question

Watson, Emma, April 15, 1990, Paris, 6:00 a.m., www.emmawatsonofficial.com

Wayne, John, May 26, 1907, Winterset, IA, time unknown, ADB#X

Weber, Jeanne, October 7, 1874, Keritry, France, 11:00 a.m., ADB#AA

Weil, Simone, February 3, 1909, Paris, France, 5:00 a.m., ADB#AA

Wilde, Kim, November 18, 1960, London, England, 8:00 p.m., ADB#A

Wilder, Gene, ne Silberman, June 11, 1933, Milwaukee, Wisconsin, 3:50 a.m., ADB#AA

William, Prince of England, June 21, 1982, London, England, 9:03 p.m., ADB#AA

Williams, Robin, July 21, 1951, Chicago, IL, 1:34 p.m., ADB#AA

Williams, Serena, September 26, 1981, Saganaw, MI, 8:28 p.m., ADB#AA

Williams, Venus, June 17, 1980, Lynwood, CA, 2:12 p.m., ADB#AA

Williamson, Marianne, July 8, 1952, Houston, TX, 3:53 a.m., ADB#A

Winfrey, Oprah, January 29, 1954, Kosciusko, MS, 4:30 a.m. CST, ADB#A

Winslet, Kate, October 5, 9175, Reading, England, 7:15 a.m. GMD, Astrotheme

Woodman, Marion, August 15, 1928, London, Canada, 2:30 p.m., given in person to author

Woods, Tiger, ne Eldrick, December 30, 1975, Long Beach, CA, 10:50 p.m., ADB#AA

Woodstock Festival, August 15, 1969, Bethel, NY, 5:07 p.m., www.woodstock69.com/
woodstock_stats.htm

Woolf, Virginia, January 25, 1882, London, England, 12:15 p.m. GMT, ADB#A

Wright, Frank Lloyd, June 8, 1867, Richland Center, WI, 5:00 p.m., ADB#A

Wyman, Bill, October 24, 1936, Lewisham, England, 11:25 p.m., ADB#B

Young, Jesse Colin, November 22, 1941, New York, 12:00 p.m., ADB#A

Young, Neil, November 12, 1945, Toronto, Canada, 2-6:00 a.m., ADB#C, time uncertain

Bibliography

Anderson, Sherry Ruth and Hopkins, Patricia. (1992). *The Feminine Face of God*. New York: Bantam Books.

Bachmann, Verena. (1997). "Lilith: The Great Goddess in Everyday Life." taped lecture, Aquarian Revelation Conference.

Baring, Anne and Cashford, Jules. (1991). *The Myth of the Goddess: Evolution of an Image.* London: Viking Arkana.

Begg, Ean. (1985). *The Cult of the Black Virgin.* London: Penguin Books.

Beriault, Marc. (2000). *La Lune Noire: Vers l'Autonomie de l'Etre*. Monaco: Editions du Rocher.

Berry, Jean-Francois. (2007). *Le Chant de la Resonance*. Saint-Andre-de-Sangonis, France: Editions Berganel.

Bige, Luc. (2004). *La Lune Noire, un vertige d'Absolu*. Paris: Les Editions de Janus.

Bloch, Ariel and Bloch, Chana. (1998). *The Song of Songs: A New Translation*. Berkeley, California: University of Callifornia Press.

Bloch, Doug and George, Demetra. (1990). *Asteroid Goddesses*. San Diego, California: ACS Publications.

Borysenko, Joan. (1999). *A Woman's Journey to God: Finding the Feminine Path*. New York: Riverhead Books.

Brady, Bernadette.(1999). *Brady's Book of Fixed Stars*. York Beach, Maine: Samuel Weiser.

Carotenuto, Aldo. (1989). *Eros and Pathos: Shades of Love and Suffering*. Toronto: Inner City Books.

De Gravelaine, Joelle. (1985). *Le Retour de Lilith: La Lune Noir*. Paris: L'Espace Bleu.

Galland, China. (1990). *Longing for Darkness*. New York: Penguin.

George, Demetra. (1992). *Mysteries of the Dark Moon*. HarperSanFrancisco.

—(2008). *Finding Our Way Through the Dark*. Tempe, Arizona: American Federation of Astrologers.

Gottlieb, Lynn. (1995). *She Who Dwells Within: A feminine vision of a renewed Judaism*. HarperSanFrancisco.

Green, Miranda. (1996). *Celtic Goddesses: Warriors, Virgins and Mothers*. New York: George Braziller.

Gustafson, Fred. (1990). *The Black Madonna*. Boston: Sigo Press.

Harding, Elizabeth U. (1993. *Kali: The Black Goddess of Dakshineswar*. York Beach, Maine: Nicolas-Hays.

Hillman, James. (1995). *Anima: An anatomy of a personified notion*. Dallas, Texas: Spring Publications.

Hunter, Kelley. (2009). *Living Lilith: Four Dimensions of the Cosmic Feminine*. Bournemouth, UK: The Wessex Astrologer, Inc.

Hurwitz, Siegmund. (1999). *Lilith—The First Eve*. Einsiedeln, Switzerland: Daimon-Verlag.

Jay, Delphine. (1981). *Interpreting Lilith*. Tempe, Arizona: American Federation of Astrologers.

—(1983). *The Lilith Ephemeris, 1900-2000 AD*. Tempe, Arizona: American Federation of Astrologers.

Johnsen, Linda, "The Legend of Lalita," *Yoga International*, Issue no. 42, June/July 1998, pp. 24-31.

Koltuv, Barbara. (1986). *The Book of Lilith*. York Beach, Maine: Nicolas-Hays.

Lehman, J. Lee and Morrison, Al. H. (1980). *Ephemeris of Lilith*. New York: CAO Times.

MacDonald, George. (1964). *Phantastes and Lilith*. Grand Rapids, Michigan: Wm. B. Eerdsmans.

Mann, A.T. and Lyle, Jane. (1995). *Sacred Sexuality*. Rockport, Massachusetts: Element Books.

Markale, Jean. (1986). *Women of the Celts*. Rochester, Vermont: Inner Traditions.

Editions St. Michel. (1993). *The New International Ephemerides 1900-2050*. St.-Michel-de-Boulogne, France.

May, Rollo, *Love and Will*. (1969). New York: Dell Publishing.

Pagels, Elaine. (1981). *The Gnostic Gospels*. New York: Random House.

Pereira, Filomena Maria. (1998). *Lilith: The Edge of Forever*. Las Colinas, Texas: Ide House.

Perera, Sylvia. (1981). *Descent to the Goddess*. Toronto: Inner City Books.

Redgrove, Peter. (1987). *The Black Goddess and the Unseen Real*. New York: Grove Press.

Rilke, R.M., *Book of Hours: Love Poems to God*. (1996). Trans. by Anita Barrows and Joanna Macy, New York: Riverhead Books.

Rossen, Robert. (1964). *Lilith*. Centaur Films.

Salamanca, J.R. (1961). *Lilith*. New York: Simon and Schuster.

Sardello, Robert. (1992). *Facing the World with Soul*. Hudson, NY: Lindesfarne.

Scharp, Susanne. (1997). *Sophia: Aspects of the Divine Feminine Past and Present*. York Beach, Maine: Nicolas-Hays.

Schmed-Kik, Liduina. "The Dark Luminaries," *The Mountain Astrologer*, June, 1996, pp. 52-57.

Shaw, Bernard. (1927). *Back to Methuselah: A metabiological pentateuch*. London: Constable and Company.

Sterling, George. (1926). *Lilith, A Dramatic Poem,* New York:The Macmillan Company.

Simmer-Brown, Judith. (2001). *Dakini's warm breath: The feminine principle in Tibetan Buddhism*. Boston: Shambhala.

Sterling, George. (1926). *Lilith, A Dramatic Poem*. New York: The Macmillan Company.

Svoboda, Robert, "When Kali Comes to Call," *Yoga International*, Issue no. 42, June/July 1998, pp. 19-23.

Wolkstein, Diane and Kramer, Samuel. (1983). *Inanna, Queen of Heaven and Earth*. New York: Harper and Row.

www.astro.com

www.astrotheme.com

www.aurea.org

www.expreso.co.cr/centaurs/blackmoon/barycentric.html//scrennu.com/astrology/

Ephemerides
Sign Changes for the Mean Black Moon 1921-2050

Aquarius	April 10, 1921
Pisces	January 4, 1922
Aries	October 1, 1922
Taurus	June 28, 1923
Gemini	March 22, 1924
Cancer	December 16, 1924
Leo	September 12, 1925
Virgo	June 9, 1926
Libra	March 5, 1927
Scorpio	November 28, 1927
Sagittarius	August 24, 1928
Capricorn	May 22, 1929
Aquarius	February 14, 1930
Pisces	November 9, 1930
Aries	August 6, 1931
Taurus	May 3, 1932
Gemini	January 26, 1933
Cancer	October 21, 1933
Leo	July 18, 1934
Virgo	April 15, 1935

Libra	January 8, 1936
Scorpio	October 2, 1936
Sagittarius	June 29, 1937
Capricorn	March 27, 1938
Aquarius	December 21, 1938
Pisces	September 15, 1939
Aries	June 10, 1940
Taurus	March 8, 1941
Gemini	December 2, 1941
Cancer	August 27, 1942
Leo	May 23, 1943
Virgo	February 18, 1944
Libra	November 13, 1944
Scorpio	August 8, 1945
Sagittarius	May 4, 1946
Capricorn	January 30, 1947
Aquarius	October 27, 1947
Pisces	July 20, 1948
Aries	April 15, 1949
Taurus	January 11, 1950
Gemini	October 8, 1950
Cancer	July 3, 1951
Leo	March 28, 1952
Virgo	December 23, 1952
Libra	September 19, 1953
Scorpio	June 14, 1954
Sagittarius	March 10, 1955
Capricorn	December 5, 1955
Aquarius	August 31, 1956
Pisces	May 26, 1957
Aries	February 19, 1958
Taurus	November 16, 1958
Gemini	August 13, 1959

Cancer	May 7, 1960
Leo	January 31, 1961
Virgo	October 29, 1961
Libra	July 26, 1962
Scorpio	April 20, 1963
Sagittarius	January 13, 1964
Capricorn	October 10, 1964
Aquarius	July 7, 1965
Pisces	April 1, 1966
Aries	December 25, 1966
Taurus	September 22, 1967
Gemini	June 18, 1968
Cancer	March 13, 1969
Leo	December 6, 1969
Virgo	September 3, 1970
Libra	May 31, 1971
Scorpio	February 23, 1972
Sagittarius	November 18, 1972
Capricorn	August 15, 1973
Aquarius	May 12, 1974
Pisces	February 5, 1975
Aries	October 31, 1975
Taurus	July 27, 1976
Gemini	April 23, 1977
Cancer	January 17, 1978
Leo	October 12, 1978
Virgo	July 9, 1979
Libra	April 4, 1980
Scorpio	December 29, 1980
Sagittarius	September 23, 1981
Capricorn	June 20, 1982
Aquarius	March 18, 1983
Pisces	December 12, 1983
Aries	September 4, 1984

Taurus	June 1, 1985
Gemini	February 27, 1986
Cancer	November 23, 1986
Leo	August 18, 1987
Virgo	May 13, 1988
Libra	February 8, 1989
Scorpio	November 4, 1989
Sagittarius	July 30 1990
Capricorn	April 25, 1991
Aquarius	January 21, 1992
Pisces	October 16, 1992
Aries	July 11, 1993
Taurus	April 6, 1994
Gemini	January 2, 1995
Cancer	September 28, 1995
Leo	June 22, 1996
Virgo	March 18, 1997
Libra	December 14, 1997
Scorpio	September 10, 1998
Sagittarius	June 5, 1999
Capricorn	February 28, 2000
Aquarius	November 25, 2000
Pisces	August 22, 2001
Aries	May 17, 2002
Taurus	February 10, 2003
Gemini	November 7, 2003
Cancer	August 3, 2004
Leo	April 28, 2005
Virgo	January 22, 2006
Libra	October 19, 2006
Scorpio	July 16, 2007
Sagittarius	April 9, 2008
Capricorn	January 3, 2009
Aquarius	September 30, 2009

Pisces	June 27, 2010
Aries	March 23, 2011
Taurus	December 16, 2011
Gemini	September 11, 2012
Cancer	June 9, 2013
Leo	March 4, 2014
Virgo	November 27, 2014
Libra	August 24, 2015
Scorpio	May 21, 2016
Sagittarius	February 13, 2017
Capricorn	November 8, 2017
Aquarius	August 5, 2018
Pisces	May 3, 2019
Aries	January 27, 2020
Taurus	October 21, 2020
Gemini	July 18, 2021
Cancer	April 14, 2022
Leo	January 8, 2023
Virgo	October 3, 2023
Libra	June 29, 2024
Scorpio	March 26, 2025
Sagittarius	December 20, 2025
Capricorn	September 14, 2026
Aquarius	June 11, 2027
Pisces	March 7, 2028
Aries	December 1, 2028
Taurus	August 26, 2029
Gemini	May 23, 2030
Cancer	February 17, 2031
Leo	November 14, 2031
Virgo	August 7, 2032
Libra	May 4, 2033
Scorpio	January 30, 2034
Sagittarius	October 26, 2034

Capricorn	July 21, 2035
Aquarius	April 15, 2036
Pisces	January 11, 2037
Aries	October 7, 2037
Taurus	July 2, 2038
Gemini	March 28, 2039
Cancer	December 24, 2039
Leo	September 18, 2040
Virgo	June 13, 2041
Libra	March 9, 2042
Scorpio	December 5, 2042
Sagittarius	August 31, 2043
Capricorn	May 25, 2044
Aquarius	February 18, 2045
Pisces	November 16, 2045
Aries	August 13, 2046
Taurus	May 8, 2047
Gemini	January 31, 2048
Cancer	October 28, 2048
Leo	July 25, 2049
Virgo	April 19, 2050

Lightning Source UK Ltd.
Milton Keynes UK
UKHW030633200222
398943UK00004B/275